John William Dawson

The Geological History of Plants

John William Dawson

The Geological History of Plants

ISBN/EAN: 9783337337957

Printed in Europe, USA, Canada, Australia, Japan

Cover: Foto ©berggeist007 / pixelio.de

More available books at **www.hansebooks.com**

THE

GEOLOGICAL HISTORY

OF PLANTS

BY

SIR J. WILLIAM DAWSON

C.M.G., LL.D., F.R.S., ETC.

WITH ILLUSTRATIONS

LONDON

KEGAN PAUL, TRENCH & CO., 1, PATERNOSTER SQUARE

1888

PREFACE.

THE object of this work is to give, in a connected form, a summary of the development of the vegetable kingdom in geological time.

To the geologist and botanist the subject is one of importance with reference to their special pursuits, and one on which it has not been easy to find any convenient manual of information. It is hoped that its treatment in the present volume will also be found sufficiently simple and popular to be attractive to the general reader.

In a work of so limited dimensions, detailed descriptions cannot be given, except occasionally by way of illustration; but references to authorities will be made in foot-notes, and certain details, which may be useful to collectors and students, will be placed in notes appended to the chapters, so as not to encumber the text.

The illustrations of this work are for the most part original; but some of them have previously appeared in special papers of the author.

J. W. D.

February, 1888.

CONTENTS.

CHAPTER VII.

CHAPTER VIII.

APPENDIX.

LIST OF ILLUSTRATIONS.

THE

GEOLOGICAL HISTORY OF PLANTS.

CHAPTER I.

PRELIMINARY IDEAS OF GEOLOGICAL CHRONOLOGY AND OF THE CLASSIFICATION OF PLANTS.

THE knowledge of fossil plants and of the history of the vegetable kingdom has, until recently, been so fragmentary that it seemed hopeless to attempt a detailed treatment of the subject of this little book. Our stores of knowledge have, however, been rapidly accumulating in recent years, and we have now arrived at a stage when every new discovery serves to render useful and intelligible a vast number of facts previously fragmentary and of uncertain import.

The writer of this work, born in a district rich in fossil plants, began to collect and work at these as a boy, in connection with botanical and geological pursuits. He has thus been engaged in the study of fossil plants for nearly half a century, and, while he has published much on the subject, has endeavoured carefully to keep within the sphere of ascertained facts, and has made it a specialty to collect, as far as possible, what has been published by others. He has also enjoyed opportunities of correspondence or personal intercourse with most of

B

the more eminent workers in the subject.   Now, in the evening of his days, he thinks it right to endeavour to place before the world a summary of facts and of his own matured conclusions—feeling, however, that nothing can be final in this matter ; and that he can only hope to sketch the present aspect of the subject, and to point the way to new developments, which must go on long after he shall have passed away.

The subject is one which has the disadvantage of pre-supposing some knowledge of the geological history of the earth, and of the classification and structures of modern plants ; and in order that all who may please to read the following pages may be placed, as nearly as possible, on the same level, this introductory chapter will be devoted to a short statement of the general facts of geological chronology, and of the natural divisions of the vegetable kingdom in their relations to that chronology.

The crust of the earth, as we somewhat modestly term that portion of its outer shell which is open to our observation, consists of many beds of rock superimposed on each other, and which must have been deposited successively, beginning with the lowest.   This is proved by the structure of the beds themselves, by the markings on their surfaces, and by the remains of animals and plants which they contain ; all these appearances indicating that each successive bed must have been the surface before it was covered by the next.

As these beds of rock were mostly formed under water, and of material derived from the waste of land, they are not universal, but occur in those places where there were extensive areas of water receiving detritus from the land. Further, as the distinction of land and water arises primarily from the shrinkage of the mass of the earth, and from the consequent collapse of the crust in some places and ridging of it up in others, it follows that there have, from the earliest geological periods, been deep ocean-basins, ridges of elevated land, and broad plateaus intervening between the ridges, and which were at some times under water, and at other times land, with many intermediate phases. The settlement and crumpling of the crust were not continuous, but took place at intervals ; and each such settlement produced not only a ridging up along certain lines, but also an emergence of the plains or plateaus. Thus at all times there have been ridges of folded rock constituting mountain-ranges, flat expansions of continental plateau, sometimes dry and sometimes submerged, and deep ocean-basins, never except in some of their shallower portions elevated into land.

By the study of the successive beds, more especially of those deposited in the times of continental submergence, we obtain a table of geological chronology which expresses the several stages of the formation of the earth's crust, from that early time when a solid shell first formed on our nascent planet to the present day. By collecting the fossil remains embedded in the several layers and placing these in chronological order, we obtain in like manner histories of animal and plant life parallel to the physical changes indicated by the beds themselves. The facts as to the sequence we obtain from the study of exposures in cliffs, cuttings, quarries, and mines ; and by correlating these local sections in a great number of places, we obtain our general table of succession ; though it is to be observed that in some single exposures or series of exposures, like those in the great cañons of Colorado, or on the coasts of Great Britain, we can often in one locality see nearly the whole sequence of beds. Let us observe here also that, though we can trace these series of deposits over the whole of the surfaces of the continents, yet if the series could be seen in one spot, say in one shaft sunk through the whole thickness of the earth's crust, this would be sufficient for our purpose, so far as the history of life is concerned.

The evidence is similar to that obtained by Schliemann on the site of Troy, where, in digging through successive layers of *débris,* he found the objects deposited by successive occupants of the site, from the time of the Roman Empire back to the earliest tribes, whose flint weapons and the ashes of their fires rest on the original surface of the ground.

Let us now tabulate the whole geological succession with the history of animals and plants associated with it :

ANIMALS.		SYSTEMS OF FORMATIONS.	PLANTS.
Age of Man and Mammalia.	Kainozoic.	Modern, Pleistocene, Pliocene, Miocene, Eocene.	Angiosperms and Palms dominant.
Age of Reptiles.	Mesozoic.	Cretaceous, Jurassic, Triassic.	Cycads and Pines dominant.
Age of Amphibians and Fishes. Age of Invertebrates.	Palæozoic.	Permian, Carboniferous, Erian, Silurian, Ordovician, Cambrian, Huronian (Upper).	Acrogens and Gymnosperms dominant.
Age of Protozoa.	Eozoic.	Huronian (Lower), Upper Laurentian, Middle Laurentian, Lower Laurentian.	Protogens and Algæ.

It will be observed, since only the latest of the systems of formations in this table belongs to the period of human history, that the whole lapse of time embraced in the table must be enormous.   If we suppose the modern period to have continued for say ten thousand years, and each of the others to have been equal to it, we shall require two hundred thousand years for the whole.   There is, however, reason to believe, from the great thickness of the formations and the slowness of the deposition of many

of them in the older systems, that they must have required vastly greater time. Taking these criteria into
account, it has been estimated that the time-ratios for
the first three great ages may be as one for the Kainozoic
to three for the Mesozoic and twelve for the Palæozoic,
with as much for the Eozoic as for the Palæozoic. This is
Dana's estimate. Another, by Hull and Houghton, gives
the following ratios : Azoic, 34·3 per cent. ; Palæozoic,
42·5 per cent. ; Mesozoic and Kainozoic, 23·2 per cent.
It is further held that the modern period is much shorter
than the other periods of the Kainozoic, so that our
geological table may have to be measured by millions of
years instead of thousands.

We cannot, however, attach any certain and definite
value in years to geological time, but must content ourselves with the general statement that it has been vastly
long in comparison to that covered by human history.

Bearing in mind this great duration of geological time,
and the fact that it probably extends from a period when
the earth was intensely heated, its crust thin, and its continents as yet unformed, it will be evident that the conditions of life in the earlier geologic periods may have
been very different from those which obtained later.
When we further take into account the vicissitudes of
land and water which have occurred, we shall see that
such changes must have produced very great differences
of climate. The warm equatorial waters have in all
periods, as superficial oceanic currents, been main agents
in the diffusion of heat over the surface of the earth, and
their distribution to north and south must have been
determined mainly by the extent and direction of land,
though it may also have been modified by the changes in
the astronomical relations and period of the earth, and
the form of its orbit.* We know by the evidence of

---

* Croll, "Climate and Time."

fossil plants that changes of this kind have occurred so great as, on the one hand, to permit the plants of warm temperate regions to exist within the Arctic Circle ; and, on the other, to drive these plants into the tropics and to replace them by Arctic forms. It is evident also that in those periods when the continental areas were largely submerged, there might be an excessive amount of moisture in the atmosphere, greatly modifying the climate, in so far as plants are concerned.

Let us now consider the history of the vegetable kingdom as indicated in the few notes in the right-hand column of the table.

The most general subdivision of plants is into the two great series of Cryptogams, or those which have no manifest flowers, and produce minute spores instead of seeds ; and Phænogams, or those which possess flowers and produce seeds containing an embryo of the future plant.

The Cryptogams may be subdivided into the following three groups :

1. *Thallogens*, cellular plants not distinctly distinguishable into stem and leaf.   These are the Fungi, the Lichens, and the Algæ, or sea-weeds.

2. *Anogens*, having stem and foliage, but wholly cellular.   These are the Mosses and Liverworts.

3. *Acrogens*, which have long tubular fibres as well as cells in their composition, and thus have the capacity of attaining a more considerable magnitude.   These are the Ferns (*Filices*), the Mare's-tails (*Equisetaceæ*), and the Club-mosses (*Lycopodiaceæ*), and a curious little group of aquatic plants called Rhizocarps (*Rhizocarpeæ*).

The Phænogams are all vascular, but they differ much in the simplicity or complexity of their flowers or seeds. On this ground they admit of a twofold division :

1. *Gymnosperms*, or those which bear naked seeds not enclosed in fruits.   They are the Pines and their allies, and the Cycads.

2. *Angiosperms*, which produce true fruits enclosing the seeds. In this group there are two well-marked subdivisions differing in the structure of the seed and stem. They are the *Endogens*, or inside growers, with seeds having one seed-leaf only, as the grasses and the palms; and the *Exogens*, having outside-growing woody stems, and seeds with two seed-leaves. Most of the ordinary forest-trees of temperate climates belong to this group.

On referring to the geological table, it will be seen that there is a certain rough correspondence between the order of rank of plants and the order of their appearance in time. The oldest plants that we certainly know are Algæ, and with these there are plants apparently with the structures of Thallophytes but the habit of trees, and which, for want of a better name, I may call *Protogens*. Plants akin to the Rhizocarps also appear very early. Next in order we find forests in which gigantic Ferns and Lycopods and Mare's-tails predominate, and are associated with pines. Succeeding these we have a reign of Gymnosperms, and in the later formations we find the higher Phænogams dominant. Thus there is an advance in elevation and complexity along with the advance in geological time, but connected with the remarkable fact that in earlier times low groups attain to an elevation unexampled in later times, when their places are occupied with plants of higher type.

It is this historical development that we have to trace in the following pages, and it will be the most simple and at the same time the most instructive method to consider it in the order of time.

CHAPTER II.

OLDEST of all the formations known to geologists, and representing perhaps the earliest rocks produced after our earth had ceased to be a molten mass, are the hard, crystalline, and much-contorted rocks named by the late Sir W. E. Logan Laurentian, and which are largely developed in the northern parts of North America and Europe, and in many other regions. So numerous and extensive, indeed, are the exposures of these rocks, that we have good reason to believe that they underlie all the other formations of our continents, and are even world-wide in their distribution. In the lower part of this great system of rocks which, in some places at least, is thirty thousand feet in thickness, we find no traces of the existence of any living thing on the earth. But, in the middle portion of the Laurentian, rocks are found which indicate that there were already land and water, and that the waters and possibly the land were already tenanted by living beings. The great beds of limestone which exist in this part of the system furnish one indication of this. In the later geological formations the limestones are mostly organic—that is, they consist of accumulated remains of shells, corals, and other hard parts of marine animals, which are composed of calcium carbonate, which the animals obtain directly from their food, and indirectly from the calcareous matter dissolved in the sea-water. In like

manner great beds of iron-ore exist in the Laurentian ;
but in later formations the determining cause of the
accumulation of such beds is the partial deoxidation and
solution of the peroxide of iron by the agency of organic
matter. Besides this, certain forms known as *Eozoon
Canadense* have been recognised in the Laurentian limestones, which indicate the presence at least of one of the
lower types of marine animals. Where animal life is, we
may fairly infer the existence of vegetable life as well,
since the plant is the only producer of food for the animal. But we are not left merely to this inference. Great
quantities of carbon or charcoal in the form of the substance known as graphite or plumbago exist in the
Laurentian. Now, in more recent formations we have
deposits of coal and bituminous matter, and we know
that these have arisen from the accumulation and slow
putrefaction of masses of vegetable matter. Further, in
places where igneous action has affected the beds, we
find that ordinary coal has been changed into anthracite
and graphite, that bituminous shales have been converted
into graphitic shales, and that cracks filled with soft
bituminous matter have ultimately become changed into
veins of graphite. When, therefore, we find in the Laurentian thick beds of graphite and beds of limestone
charged with detached grains and crystals of this substance, and graphitic gneisses and schists and veins of
graphite traversing the beds, we recognise the same
phenomena that are apparent in later formations containing vegetable *débris*.

The carbon thus occurring in the Laurentian is not
to be regarded as exceptional or rare, but is widely distributed and of large amount. In Canada more especially
the deposits are very considerable.

The graphite of the Laurentian of Canada occurs both
in beds and in veins, and in such a manner as to show
that its origin and deposition are contemporaneous with

those of the containing rock.   Sir William Logan states *
that "the deposits of plumbago generally occur in the
limestones or in their immediate vicinity, and granular
varieties of the rock often contain large crystalline plates
of plumbago.   At other times this mineral is so finely
disseminated as to give a bluish-grey colour to the limestone, and the distribution of bands thus coloured seems
to mark the stratification of the rock."   He further
states : "The plumbago is not confined to the limestones ; large crystalline scales of it are occasionally disseminated in pyroxene rock, and sometimes in quartzite
and in feldspathic rocks, or even in magnetic oxide of
iron."   In addition to these bedded forms, there are also
true veins in which graphite occurs associated with calcite, quartz, orthoclase, or pyroxene, and either in disseminated scales, in detached masses, or in bands or layers
"separated from each other and from the wall-rock by
feldspar, pyroxene, and quartz."   Dr. Hunt also men-
tions the occurrence of finely granular varieties, and of
that peculiarly waved and corrugated variety simulating
fossil wood, though really a mere form of laminated
structure, which also occurs at Warrensburg, New York,
and at the Marinski mine in Siberia.   Many of the veins
are not true fissures, but rather constitute a network of
shrinkage cracks or segregation veins traversing in countless numbers the containing rock, and most irregular in
their dimensions, so that they often resemble strings of
nodular masses.   It is most probable that the graphite of
the veins was originally introduced as a liquid or plastic
hydrocarbon ; but in whatever way introduced, the character of the veins indicates that in the case of the greater
number of them the carbonaceous material must have
been derived from the bedded rocks traversed by these
veins, to which it bears the same relation with the veins

---

* "Geology of Canada," 1863.

of bitumen found in the bituminous shales of the Carboniferous and Silurian rocks. Nor can there be any doubt that the graphite found in the beds has been deposited along with the calcareous matter or muddy and sandy sediment of which these beds were originally composed.*

The quantity of graphite in the Lower Laurentian series is enormous. Some years ago, in the township of Buckingham, on the Ottawa River, I examined a band of limestone believed to be a continuation of that described by Sir W. E. Logan as the Green Lake limestone. It was estimated to amount, with some thin interstratified bands of gneiss, to a thickness of six hundred feet or more, and was found to be filled with disseminated crystals of graphite and veins of the mineral to such an extent as to constitute in some places one-fourth of the whole ; and, making every allowance for the poorer portions, this band cannot contain in all a less vertical thickness of pure graphite than from twenty to thirty feet. In the adjoining township of Lochaber Sir W. E. Logan notices a band from twenty-five to thirty feet thick, reticulated with graphite veins to such an extent as to be mined with profit for the mineral. At another place in the same district a bed of graphite from ten to twelve feet thick, and yielding 20 per cent. of the pure material, is worked. As it appears in the excavation made by the quarrymen, it resembled a bed of coal; and a block from this bed, about four feet thick, was a prominent object in the Canadian department of the Colonial Exhibition of 1886. When it is considered that graphite occurs in similar abundance at several other horizons, in beds of limestone which have been ascertained by Sir W. E. Logan to have an aggregate thickness of thirty-five hundred feet, it is

---

* Paper by the author on Laurentian Graphite, "Journal of London Geological Society," 1876.

scarcely an exaggeration to maintain that the quantity of carbon in the Laurentian is equal to that in similar areas of the Carboniferous system.   It is also to be observed that an immense area in Canada appears to be occupied by these graphitic and *Eozoon* limestones, and that rich graphitic deposits exist in the continuation of this system in the State of New York, while in rocks believed to be of this age near St. John, New Brunswick, there is a very thick bed of graphitic limestone, and associated with it three regular beds of graphite, having an aggregate thickness of about five feet.*

It may fairly be assumed that in the present world, and in those geological periods with whose organic remains we are more familiar than with those of the Laurentian, there is no other source of unoxidized carbon in rocks than that furnished by organic matter, and that this has obtained its carbon in all cases, in the first instance, from the deoxidation of carbonic acid by living plants.   No other source of carbon can, I believe, be imagined in the Laurentian period.   We may, however, suppose either that the graphitic matter of the Laurentian has been accumulated in beds like those of coal, or that it has consisted of diffused bituminous matter similar to that in more modern bituminous shales and bituminous and oil-bearing limestones.   The beds of graphite near St. John, some of those in the gneiss at Ticonderoga in New York, and at Lochaber and Buckingham, and elsewhere in Canada, are so pure and regular that one might fairly compare them with the graphitic coal of Rhode Island.   These instances, however, are exceptional, and the greater part of the disseminated and vein graphite might rather be likened in its mode of occurrence to the bituminous matter in bituminous shales and limestones.

---

* Matthew in "Quarterly Journal of the Geological Society," vol. xxi., p. 423. "Acadian Geology," p. 662.

We may compare the disseminated graphite to that which we find in those districts of Canada in which Silurian and Devonian bituminous shales and limestones have been metamorphosed and converted into graphitic rocks not very dissimilar to those in the less altered portions of the Laurentian.* In like manner it seems probable that the numerous reticulating veins of graphite may have been formed by the segregation of bituminous matter into fissures and planes of least resistance, in the manner in which such veins occur in modern bituminous limestones and shales. Such bituminous veins occur in the Lower Carboniferous limestone and shale of Dorchester and Hillsborough, New Brunswick, with an arrangement very similar to that of the veins of graphite; and in the Quebec rocks of Point Levi, veins attaining to a thickness of more than a foot, are filled with a coaly matter having a transverse columnar structure, and regarded by Logan and Hunt as an altered bitumen. These palæozoic analogies would lead us to infer that the larger part of the Laurentian graphite falls under the second class of deposits above mentioned, and that, if of vegetable origin, the organic matter must have been thoroughly disintegrated and bituminised before it was changed into graphite. This would also give a probability that the vegetation implied was aquatic, or at least that it was accumulated under water.

Dr. Hunt has, however, observed an indication of terrestrial vegetation, or at least of subaërial decay, in the great beds of Laurentian iron-ore. These, if formed in the same manner as more modern deposits of this kind, would imply the reducing and solvent action of substances produced in the decay of plants. In this case such great ore-beds as that of Hull, on the Ottawa, seventy

---

* Granby, Melbourne, Owl's Head, &c., "Geology of Canada," 1863,

feet thick, or that near Newborough, two hundred feet thick,* must represent a corresponding quantity of vegetable matter which has totally disappeared. It may be added that similar demands on vegetable matter as a deoxidising agent are made by the beds and veins of metallic sulphides of the Laurentian, though some of the latter are no doubt of later date than the Laurentian rocks themselves.

It would be very desirable to confirm such conclusions as those above deduced by the evidence of actual microscopic structure. It is to be observed, however, that when, in more modern sediments, Algæ have been converted into bituminous matter, we cannot ordinarily obtain any structural evidence of the origin of such bitumen, and in the graphitic slates and limestones derived from the metamorphosis of such rocks no organic structure remains. It is true that, in certain bituminous shales and limestones of the Silurian system, shreds of organic tissue can sometimes be detected, and in some cases, as in the Lower Silurian limestone of the La Cloche Mountains in Canada, the pores of brachiopodous shells and the cells of corals have been penetrated by black bituminous matter, forming what may be regarded as natural injections, sometimes of much beauty. In correspondence with this, while in some Laurentian graphitic rocks, as, for instance, in the compact graphite of Clarendon, the carbon presents a curdled appearance due to segregation, and precisely similar to that of the bitumen in more modern bituminous rocks, I can detect in the graphitic limestones occasional fibrous structures which may be remains of plants, and in some specimens vermicular lines, which I believe to be tubes of *Eozoon* penetrated by matter once bituminous, but now in the state of graphite.

---

* " Geology of Canada," 1863.

When palæozoic land-plants have been converted into graphite, they sometimes perfectly retain their structure. Mineral charcoal, with structure, exists in the graphitic coal of Rhode Island.   The fronds of ferns, with their minutest veins perfect, are preserved in the Devonian shales of St. John, in the state of graphite; and in the same formation there are trunks of Conifers (*Dadoxylon Ouangondianum*) in which the material of the cell-walls has been converted into graphite, while their cavities have been filled with calcareous spar and quartz, the finest structures being preserved quite as well as in comparatively unaltered specimens from the coal-formation.* No structures so perfect have as yet been detected in the Laurentian, though in the largest of the three graphitic beds at St. John there appear to be fibrous structures, which I believe may indicate the existence of land-plants. This graphite is composed of contorted and slickensided laminæ, much like those of some bituminous shales and coarse coals; and in these are occasional small pyritous masses which show hollow carbonaceous fibres, in some cases presenting obscure indications of lateral pores.   I regard these indications, however, as uncertain; and it is not as yet fully ascertained that these beds at St. John are on the same geological horizon with the Lower Laurentian of Canada, though they certainly underlie the Primordial series of the Acadian group, and are separated from it by beds having the character of the Huronian.

There is thus no absolute impossibility that distinct organic tissues may be found in the Laurentian graphite, if formed from land-plants, more especially if any plants existed at that time having true woody or vascular tissues; but it cannot with certainty be affirmed that such tissues

---

* " Acadian Geology," p. 535.   In calcified specimens the structures remain in the graphite after decalcification by an acid.

have been found. It is possible, however, that in the Laurentian period the vegetation of the land may have consisted wholly of cellular plants, as, for example, mosses and lichens; and if so, there would be comparatively little hope of the distinct preservation of their forms or tissues, or of our being able to distinguish the remains of land-plants from those of Algæ.

We may sum up these facts and considerations in the following statements : First, that somewhat obscure traces of organic structure can be detected in the Laurentian graphite ; secondly, that the general arrangement and microscopic structure of the substance corresponds with that of the carbonaceous and bituminous matters in marine formations of more modern date ; thirdly, that if the Laurentian graphite has been derived from vegetable matter, it has only undergone a metamorphosis similar in kind to that which organic matter in metamorphosed sediments of later age has experienced ; fourthly, that the association of the graphitic matter with organic limestone, beds of iron-ore, and metallic sulphides greatly strengthens the probability of its vegetable origin ; fifthly, that when we consider the immense thickness and extent of the Eozoonal and graphitic limestones and iron-ore deposits of the Laurentian, if we admit the organic origin of the limestone and graphite, we must be prepared to believe that the life of that early period, though it may have existed under low forms, was most copiously developed, and that it equalled, perhaps surpassed, in its results, in the way of geological accumulation, that of any subsequent period.

Many years ago, at the meeting of the American Association in Albany, the writer was carrying into the room of the Geological Section a mass of fossil wood from the Devonian of Gaspé, when he met the late Professor Agassiz, and remarked that the specimen was the remains of a Devonian tree contemporaneous with his fishes of that age. "How I wish I could sit under its shade!" was the smiling reply of the great zoölogist; and when we think of the great accumulations of Laurentian carbon, and that we are entirely ignorant of the forms and structures of the vegetation which produced it, we can scarcely suppress a feeling of disappointment. Some things, however, we can safely infer from the facts that are known, and these it may be well to mention.

The climate and atmosphere of the Laurentian may have been well adapted for the sustenance of vegetable life. We can scarcely doubt that the internal heat of the earth still warmed the waters of the sea, and these warm waters must have diffused great quantities of mists and vapours over the land, giving a moist and equable if not a very clear atmosphere. The vast quantities of carbon dioxide afterwards sealed up in limestones and carbonaceous beds must also have still floated in the atmosphere and must have supplied abundance of the carbon, which constitutes the largest ingredient in vegetable tissues. Under these circumstances the whole world must have resembled a damp, warm greenhouse, and plants loving such an atmosphere could have grown luxuriantly. In these circumstances the lower forms of aquatic vegetation and those that love damp, warm air and wet soil would have been at home.

If we ask more particularly what kinds of plants might be expected to be introduced in such circumstances, we may obtain some information from the vegetation of the succeeding Palæozoic age, when such conditions still continued to a modified extent. In this period the club-mosses, ferns, and mare's-tails engrossed the world and grew to sizes and attained degrees of complexity of structure not known in modern times. In the previous Laurentian age something similar may have happened to Algæ, to Fungi, to Lichens, to Liverworts, and Mosses. The Algæ may have attained to gigantic dimensions, and may have even ascended out of the water in some of their forms. These comparatively simple cellular and tubular structures, now degraded to the humble position of flat lichens or soft or corky fungi, or slender cellular mosses, may have been so strengthened and modified as to constitute forest-trees. This would be quite in harmony with what is observed in the development of other plants in primitive geological times; and a little later in this history we shall see that there is evidence in the flora of the Silurian of a survival of such forms.

It may be that no geologist or botanist will ever be able to realise these dreams of the past. But, on the other hand, it is quite possible that some fortunate chance may have somewhere preserved specimens of Laurentian plants showing their structure.

In any case we have here presented to us the strange and startling fact that the remarkable arrangement of protoplasmic matter and chlorophyll, which enables the vegetable cell to perform, with the aid of solar light, the miracle of decomposing carbon dioxide and water, and forming with them woody and corky tissues, had already been introduced upon the earth. It has been well said that no amount of study of inorganic nature would ever have enabled any one to anticipate the possibility of the construction of an apparatus having the chemical powers of the living vegetable cell. Yet this most marvellous structure seems to have been introduced in the full plenitude of its powers in the Laurentian age.

Whether this early Laurentian vegetation was the means of sustaining any animal life other than marine Protozoa, we do not know. It may have existed for its own sake alone, or merely as a purifier of the atmosphere, in preparation for the future introduction of land-animals. The fact that there have existed, even in modern times, oceanic islands rich in vegetation, yet untenanted by the higher forms of animal life, prepares us to believe that such conditions may have been general or universal in the primeval times we are here considering.

If we ask to what extent the carbon extracted from the atmosphere and stored up in the earth has been, or is likely to be, useful to man, the answer must be that it is not in a state to enable it to be used as mineral fuel. It has, however, important uses in the arts, though at present the supply seems rather in excess of the demand, and it may well be that there are uses of graphite still undiscovered, and to which it will yet be applied.

Finally, it is deserving of notice that, if Laurentian graphite indicates vegetable life, it indicates this in vast profusion. That incalculable quantities of vegetable matter have been oxidised and have disappeared we may believe on the evidence of the vast beds of iron-ore ; and, in regard to that preserved as graphite, it is certain that every inch of that mineral must indicate many feet of crude vegetable matter.

It is remarkable that, in ascending from the Laurentian, we do not at first appear to advance in evidences of plant-life. The Huronian age, which succeeded the Laurentian, seems to have been a disturbed and unquiet time, and, except in certain bands of iron-ore and some dark slates coloured with carbonaceous matter, we find in it no evidence of vegetation. In the Cambrian a great subsidence of our continents began, which went on, though with local intermissions and reversals, all through the Siluro-Cambrian or Ordovician time. These times were, for this reason, remarkable for the great abundance and increase of marine animals rather than of land-plants. Still, there are some traces of land vegetation, and we may sketch first the facts of this kind which are known, and then advert to some points relating to the earlier Algæ, or sea-weeds.

An eminent Swedish geologist, Linnarsson, has described, under the name of *Eophyton*, certain impressions on old Cambrian rocks in Sweden, and which certainly present very plant-like forms. They want, however, any trace of carbonaceous matter, and seem rather to be grooves or marks cut in clay by the limbs or tails of some aquatic animal, and afterwards filled up and preserved by succeeding deposits. After examining large series of these specimens from Sweden, and from rocks of similar age in Canada, I confess that I have no faith in their vegetable nature.

The oldest plants known to me, and likely to have been of higher grade than Algæ, are specimens kindly presented to me by Dr. Alleyne Nicholson, of Aberdeen, and which he had named *Buthotrephis Harknessii** and *B. radiata*. They are from the Skiddaw rocks of Cumberland. On examining these specimens, and others subsequently collected in the same locality by Dr. G. M. Dawson, while convinced by their form and carbonaceous character that they are really plants, I am inclined to refer them not to Algæ, but probably to Rhizocarps. They consist of slender branching stems, with whorls of elongate and pointed leaves, resembling the genus *Annularia* of the coal formation. I am inclined to believe that both of Nicholson's species are parts of one plant, and for this I have proposed the generic name *Protannularia* (Fig. 1). Somewhat higher in the Siluro-Cambrian, in the Cincinnati group of America, Lesquereux has found some minute radiated leaves, referred by him to the genus *Sphenophyllum*,† which is also allied to Rhizocarps. Still more remarkable is the discovery in the same beds of a stem with rhombic areoles or leaf-bases, to which the name *Protostigma* has been given.‡ If a plant, this may

---

* " Geological Magazine," 1869.
† See figure in next chapter.
‡ *Protostigma sigillarioides*, Lesquereux.

have been allied to the club-mosses. This seems to be all that we at present know of land-vegetation in the Siluro-Cambrian. So far as the remains go, they indicate the presence of the families of Rhizocarps and of Lycopods.

If we ascend into the Upper Silurian, or Silurian proper, the evidences of land vegetation somewhat increase. In 1859 I described, in "The Journal of the Geological Society," of London, a remarkable tree from the Lower Erian of Gaspé, under the name *Prototaxites*, but for which I now prefer the name *Nematophyton*. When in London, in 1870, I obtained permission to examine certain specimens of spore-cases or seeds from the Upper Ludlow (Silurian) formation of England, and which had been described by Sir Joseph Hooker under the name *Pachytheca*. In the same slabs with these I found fragments of fossil wood identical with those of the Gaspé plant. Still later I recognised similar fragments associated also with *Pachytheca* in the Silurian of Cape Bon Ami, New Brunswick. Lastly, Dr. Hicks has discovered similar wood, and also similar

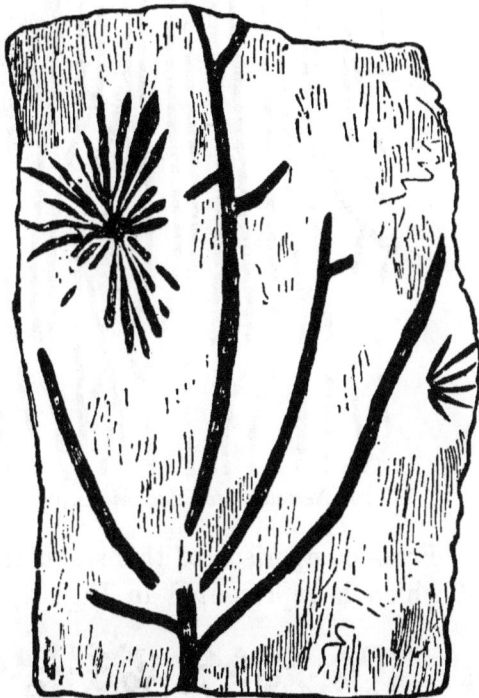

FIG. 1.—*Protannularia Harknessii* (Nicholson), a probable Rhizocarp of the Ordovician period.

fruits, in the Denbighshire grits, at the base of the Silurian.*

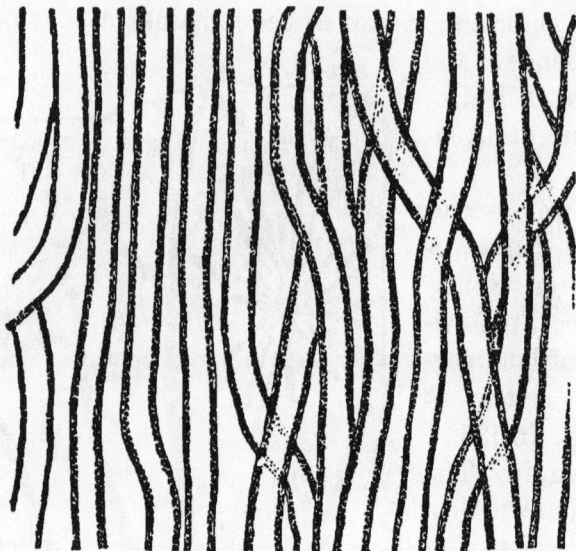

FIG. 2.—*Nematophyton Logani* (magnified).   Vertical section.

From comparison of this singular wood, the structure of which is represented in Figs. 2, 3, 4, with the *débris*

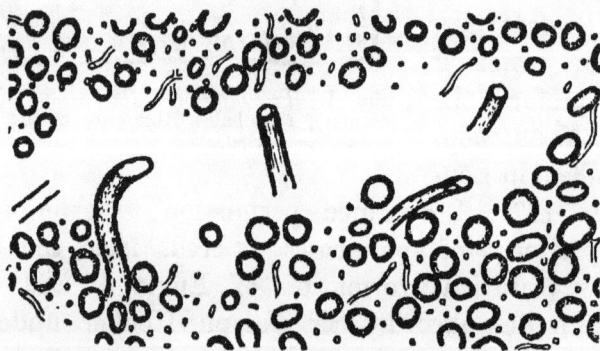

FIG. 3.—*Nematophyton Logani* (magnified).   Horizontal section, showing part of one of the radial spaces, with tubes passing into it.

of fossil taxine woods, mineralised after long maceration in water, I was inclined to regard *Prototaxites*, or, as I

---

* "Journal of the Geological Society," August, 1881.

have more recently named it, *Nematophyton*, as a primeval gymnosperm allied to those trees which Unger had described from the Erian of Thuringia, under the name *Aporoxylon*.* Later examples of more lax tissues from branches or young stems, and the elaborate examinations kindly undertaken for me by Professor Penhallow and

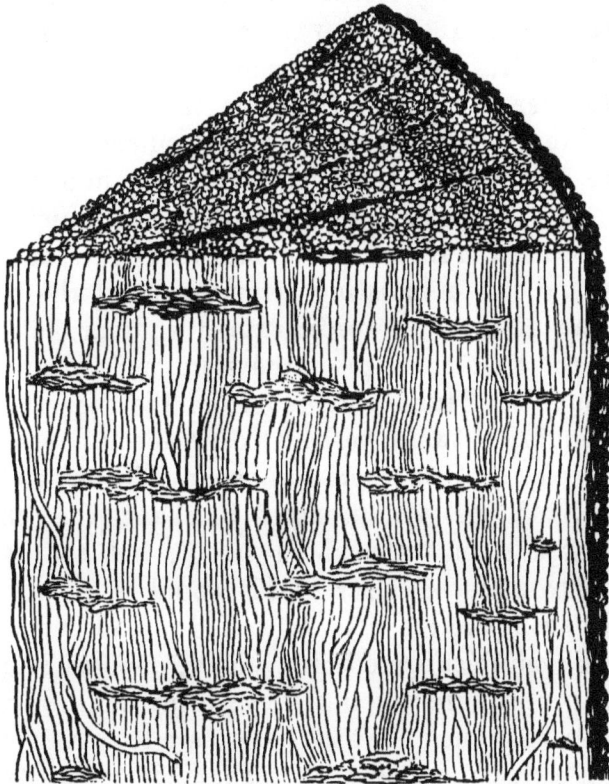

Fig. 4.—*Nematophyton Logani* (magnified).   Restoration.†

referred to in a note to this chapter, have induced me to modify this view, and to hold that the tissues of these singular trees, which seem to have existed from the be-

---

* " Palæontologie des Thuringer Waldes," 1856.

† Figs. 2, 3, and 4 are drawn from nature by Prof. Penhallow, of McGill College.

ginning of the Silurian age and to have finally disappeared in the early Erian, are altogether distinct from any form of vegetation hitherto known, and are possibly survivors of that prototypal flora to which I have already referred. They are trees of large size, with a coaly bark and large spreading roots, having the surface of the stem smooth or irregularly ribbed, but with a nodose or jointed appearance. Internally, they show a tissue of long, cylindrical tubes, traversed by a complex network of horizontal tubes thinner walled and of smaller size. The tubes are arranged in concentric zones, which, if annual rings, would in some specimens indicate an age of one hundred and fifty years. There are also radiating spaces, which I was at first disposed to regard as true medullary rays, or which at least indicate a radiating arrangement of the tissue. They now seem to be spaces extending from the centre towards the circumference of the stem, and to have contained bundles of tubes gathered from the general tissue and extending outward perhaps to organs or appendages on the surface. Carruthers has suggested a resemblance to Algæ, and has even proposed to change the name to *Nematophycus*, or "thread-sea-weed"; but the resemblance is by no means clear, and it would be quite as reasonable to compare the tissue to that of some Fungi or Lichens, or even to suppose that a plant composed of cylindrical tubes has been penetrated by the mycelium or spawn of a dry-rot fungus. But the tissues are too constant and too manifestly connected with each other to justify this last supposition. That the plant grew on land I cannot doubt, from its mode of occurrence; that it was of durable and resisting character is shown by its state of preservation; and the structure of the seeds called *Pachytheca*, with their constant association with these trees, give countenance to the belief that they are the fruit of Nematophyton. Of the foliage or fronds of these strange plants we unfortunately know nothing. They seem, how-

ever, to realise the idea of arboreal plants having structures akin to those of thallophytes, but with seeds so
large and complex that they can scarcely be regarded as
mere spores.   They should perhaps constitute a separate
class or order to which the name *Nematodendreæ* may
be given, and of which *Nematophyton* will constitute one
genus and *Aporoxylon* of Unger another.*

Another question arises as to the possible relation of
these plants to other trees known by their external forms.
The *Protostigma* of Lesquereux has already been referred
to, and Claypole has described a tree from the Clinton
group of the United States, with large ovate leaf-bases, to
which he has given the name *Glyptodendron.*† If the
markings on these plants are really leaf-bases, they can
scarcely have been connected with *Nematophyton*, because
that tree shows no such surface-markings, though, as we
have seen, it had bundles of tubes passing diagonally to
the surface.   These plants were more probably trees with
an axis of barred vessels and thick, cellular bark, like the
Lepidodendron of later periods, to be noticed in the sequel.
Dr. Hicks has also described from the same series of beds
which afforded the fragments of Nematophyton certain
carbonised dichotomous stems, which he has named *Berwynia*.   It is just possible that these plants may have
belonged to the Nematodendreæ.   The thick and dense
coaly matter which they show resembles the bark of these
trees, the longitudinal striation in some of them may
represent the fibrous structure, and the lateral projections
which have been compared to leaves or leaf-bases may
correspond with the superficial eminences of *Nematophyton*, and the spirally arranged punctures which it shows
on its surface.   In this case I should be disposed to re-

---

* See report by the author on " Erian Flora of Canada," 1871 and
1882, for full description of these fossils.
    † " American Journal of Science," 1878.

gard the supposed stigmaria-like roots as really stems, and the supposed rootlets as short, spine-like rudimentary leaves. All such comparisons must, however, in the mean time be regarded as conjectural. We seem, however, to have here a type of tree very dissimilar to any even of the later Palæozoic age, which existed throughout the Silurian, and probably further back, which ceased to exist early in the Erian age, and before the appearance of the ordinary coniferous and lepidodendroid trees. May it not have been a survivor of an old arboreal flora extending back even to the Laurentian itself?

Multitudes of markings occurring on the surfaces of the older rocks have been referred to the Algæ or seaweeds, and indeed this group has been a sort of refuge for the destitute to which palæontologists have been accustomed to refer any anomalous or inexplicable form which, while probably organic, could not be definitely referred to the animal kingdom. There can be no question that some of these are truly marine plants; and that plants of this kind occur in formations older than those in which we first find land-plants, and that they have continued to inhabit the sea down to the present time. It is also true that the oldest of these Algæ closely resemble in form plants of this kind still existing; and, since their simple cellular structures and soft tissues are scarcely ever preserved, their general forms are all that we can know, so that their exact resemblance to or difference from modern types can rarely be determined. For the same reasons it has proved difficult clearly to distinguish them from mere inorganic markings or the traces of animals, and the greatest divergence of opinion has occurred in recent times on these subjects, as any one can readily understand who consults the voluminous and well-illustrated memoirs of Nathorst, Williamson, Saporta, and Delgado.

The author of this work has given much attention to these remains, and has not been disposed to claim for the

vegetable kingdom so many of them as some of his contemporaries.* The considerations which seem most important in making such distinctions are the following :
1. The presence or absence of carbonaceous matter.
True Algæ not infrequently present at least a thin film of
carbon representing their organic matter, and this is the
more likely to occur in their case, as organic matters
buried in marine deposits and not exposed to atmospheric
oxidation are very likely to be preserved. 2. In the
absence of organic matter, the staining of the containing
rock, the disappearance or deoxidation of its ferruginous
colouring matter, or the presence of iron pyrite may indicate the removal of organic matter by decay. 3. When
organic matter and indications of it are altogether absent,
and form alone remains, we have to distinguish from Algæ,
trails and burrows similar to those of aquatic animals,
casts of shrinkage-cracks, water-marks, and rill-marks
widely diffused over the surfaces of beds. 4. Markings
depressed on the upper surfaces of beds, and filled with
the material of the succeeding layer, are usually mere impressions. The cases of possible exceptions to this are
very rare. On the contrary, there are not infrequently
forms in relief on the surfaces of rocks which are not
Algæ, but may be shallow burrows arched upward on top,
or castings of worms thrown up upon the surface. Sometimes, however, they may have been left by denudation
of the surrounding material, just as footprints on dry
snow remain in relief after the surrounding loose material
has been drifted away by the wind; the portion consolidated by pressure being better able to resist the denuding
agency.

The footprints from the Potsdam sandstone in Canada, for which the name *Protichnites* was proposed by

---

* "Impressions and Footprints of Aquatic Animals," "American
Journal of Science," 1873.

Owen, and which were by him referred to crustaceans probably resembling *Limulus*, were shown by the writer,

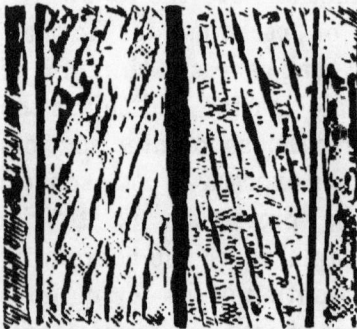

in 1862,* to correspond precisely with those of the American Limulus (*Polyphemus Occidentalis*) (Fig. 5). I proved by experiment with the modern animal that the recurring series of groups of markings were produced by the toes of the large posterior thoracic foot, the irregular scratches seen in *Protichnites lineatus* by the ordinary feet, and the central furrow by the tail. It was also shown that when the Limulus uses its swimming-feet it produces impressions of the character of those named

Fig. 5.—Trail of a modern king-crab, to illustrate imitations of plants sometimes named *Bilobites*.

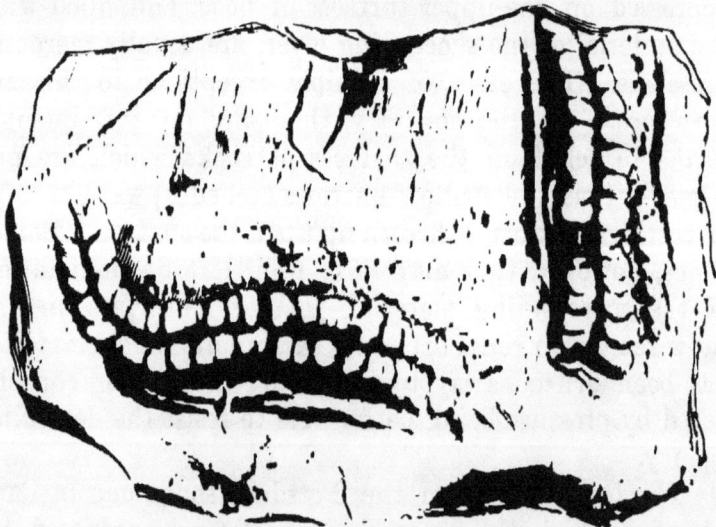

Fig. 6.—Trail of Carboniferous crustacean (*Rusichnites Acadicus*), Nova Scotia, to illustrate supposed Algæ.

---

* "Canadian Naturalist," vol. vii.

Climactichnites, from the same beds which afford *Protichnites.* The principal difference between *Protichnites* and their modern representatives is that the latter have two lateral furrows produced by the sides of the carapace, which are wanting in the former.

I subsequently applied the same explanation to several other ancient forms now known under the general name *Bilobites* (Figs. 6 and 7).*

The tuberculated impressions known as *Phymatoderma* and *Caulerpites* may, as Zeiller has shown, be made by the burrowing of the mole-

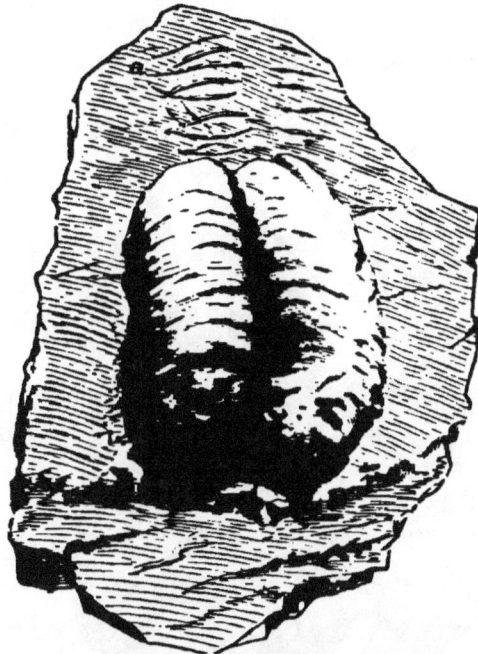

FIG. 7.—*Buthophycus (Rusichnites) Grenvillensis,* an animal burrow of the Siluro-Cambrian, probably of a crustacean. *a,* Track connected with it.

cricket, and fine examples occurring in the Clinton formation of Canada are probably the work of Crustacea. It is probable, however, that some of the later forms referred to these genera are really Algæ related to *Caulerpa,* or even branches of Conifers of the genus *Brachyphyllum.*

Nereites and *Planulites* are tracks and burrows of worms, with or without marks of setæ, and some of the

* The same *Bilobites* was originally proposed by De Kay for a bivalve shell (*Conocardium*). Its application to supposed Algæ was an error, but this is of the less consequence, as these are not true plants but only animal trails.

markings referred to *Palæochorda*, *Palæophycus*, and *Scolithus* have their places here. Many examples highly illustrative of the manner of formation of the impressions are afforded by Canadian rocks (Fig. 8).

Branching forms referred to *Licrophycus* of Billings, and some of those referred to *Buthotrephis*, Hall, as well as radiating markings referable to *Scotolithus*, *Gyrophyllites*, and *Asterophycus*, are explained by the branching burrows of worms illustrated by Nathorst and the author. *Astropolithon*, a singular radiating marking of the Canadian Cambrian,* seems to be something organic, but of what nature is uncertain (Fig. 9).

FIG. 8.—*Palæophycus Beverlyensis* (Billings), a supposed Cambrian Fucoid, but probably an animal trail.

Rhabdichnites and *Eophyton* belong to impressions explicable by the trails of drifting sea-weeds, the tail-markings of Crustacea, and the ruts ploughed by bivalve mollusks, and occurring in the Silurian, Erian, and Carboniferous rocks.† Among these are the singular bilobate forms described as *Rusophycus* by Hall, and which are probably burrows or resting-places of crustaceans. The tracks of such animals, when walking, are the jointed impressions known as *Arthrophycus* and *Crusiana*. I have shown by the mode of occurrence

---

* Supplement to "Acadian Geology."
† "Canadian Naturalist," 1864.

of these, and Nathorst has confirmed this conclusion by elaborate experiments on living animals, that these forms are really trails impressed on soft sediments by animals and mostly by crustaceans.

I agree with Dr. Williamson [*] in believing that all or nearly all the forms referred to Crossochorda of Schimper are really animal impressions allied to Nereites, and due either to worms or, as Nathorst has shown to be possible, to small crustaceans. Many impressions of this kind occur in the Silurian beds of the Clinton series in Canada and New York, and are undoubtedly mere markings.

It is worthy of note that these markings strikingly resemble the so-called *Eophyton*, described by Torell from the Primordial of Sweden, and by Billings from that of Newfoundland ; and which also occur abundantly in the Primordial of New Brunswick. After examining a series of these markings from Sweden shown to me by Mr. Carruthers in

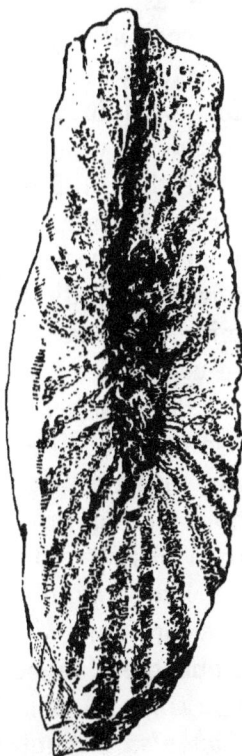

Fig. 9. — *Astropolithon Hindii*, an organism of the Lower Cambrian of Nova Scotia, possibly vegetable.

London, and also specimens from Newfoundland and a large number *in situ* at St. John, I am convinced that they cannot be plants, but must be markings of the nature of *Rhabdichnites*. This conclusion is based on the absence of carbonaceous matter, the intimate union of the markings with the surface of the stone,

---

[*] "Tracks from Yoredale Rocks," "Manchester Literary and Philosophical Society," 1885.

their indefinite forms, their want of nodes or appendages, and their markings being always of such a nature as could be produced by scratches of a sharp instrument. Since, however, fishes are yet unknown in beds of this age, they may possibly be referred to the feet or spinous tails of swimming crustaceans. Salter has already suggested this origin for some scratches of somewhat different form found in the Primordial of Great Britain. He supposed them to have been the work of species of *Hymenocaris*. These marks may, however, indicate the existence of some free-swimming animals of the Primordial seas as yet unknown to us.

Three other suggestions merit consideration in this connection. One is that Algæ and also land-plants, drifting with tides or currents, often make the most remarkable and fantastic trails. A marking of this kind has been observed by Dr. G. M. Dawson to be produced by a drifted Laminaria, and in complexity it resembled the extraordinary *Ænigmichnus multiformis* of Hitchcock from the Connecticut sandstones. Much more simple markings of this kind would suffice to give species of *Eophyton*. Another is furnished by a fact stated to the author by Prof. Morse, namely, that Lingulæ, when dislodged from their burrows, trail themselves over the bottom like worms, by means of their cirri. Colonies of these creatures, so abundant in the Primordial, may, when obliged to remove, have covered the surfaces of beds of mud with vermicular markings. The third is that the Rhabdichnite-markings resemble some of the grooves in Silurian rocks which have been referred to trails of Gasteropods, as, for instance, those from the Clinton group, described by Hall.

Another kind of markings not even organic, but altogether depending on physical causes, are the beautiful branching rill-marks produced by the oozing of water out of mud and sand-banks left by the tide, and which sometimes cover great surfaces with the most elaborate tracery, on the modern tidal shores as well as in some of the most ancient rocks. *Dendrophycus* * of Lesquereux seems to be an example of rill-mark, as well as *Aristophycus*, *Clœphycus*, and *Zygophycus*, of Miller and Dyer, from the Lower Silurian.

Rill-marks occur in very old rocks,† but are perhaps most beautifully preserved in the Carboniferous shales and argillaceous sandstones, and even more elaborately on the modern mud‑banks of the Bay of Fundy.‡ Some of these simulate ferns and fronds of Laminariæ, and others resemble roots, fucoids allied to *Buthotrephis*, or the radiating worm-burrows already referred to (Fig. 10).

Shrinkage-cracks are also abundant in some of the Carboniferous beds, and are sometimes accompanied with impressions of rain-drops. When finely reticulated they might be mistaken for the venation of leaves, and, when complicated with little rill-marks tributary to their sides, they precisely resemble the *Dictyolites* of Hall from the Medina sandstone (Fig. 11).

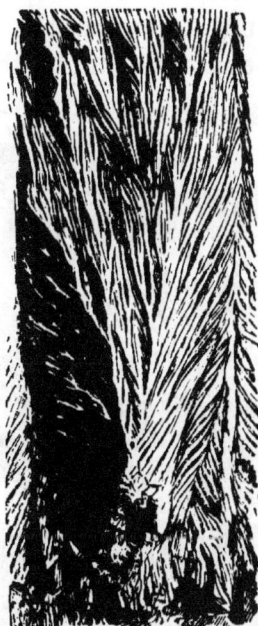

Fig. 10.—Carboniferous rill-mark (Nova Scotia), reduced, to illustrate pretended Algæ.

An entirely different kind of shrinkage-crack is that which occurs in certain carbonised and flattened plants,

---

* "Coal Flora of Pennsylvania," vol. iii., Plate 88.
† "Journal of the Geological Society," vol. xii., p. 251.
‡ "Acadian Geology," 2d ed., p. 26.

D

and which sometimes communicates to them a marvellous resemblance to the netted under surface of an exogenous leaf. Flattened stems of plants and layers of cortical matter, when carbonised, shrink in such a manner as to produce minute reticulated cracks. These become filled with mineral matter before the coaly substance has been completely consolidated. A further compression occurs, causing the coaly substance to collapse, leaving the little veins of harder mineral matter projecting. These impress their form upon the clay or shale above and below, and thus when the mass is broken open we have a carbonaceous film or thin layer covered with a network of raised lines, and corresponding minute depressed lines on the shale in contact with it. The reticulations are generally irregular, but sometimes they very closely resemble the veins of a reticulately veined leaf. One of the most curious specimens in my possession was collected by Mr. Elder in the Lower Carboniferous of Horton Bluff. The little veins which form the projecting network are in this case white calcite ; but at the surface their projecting edges are blackened with a carbonaceous film.

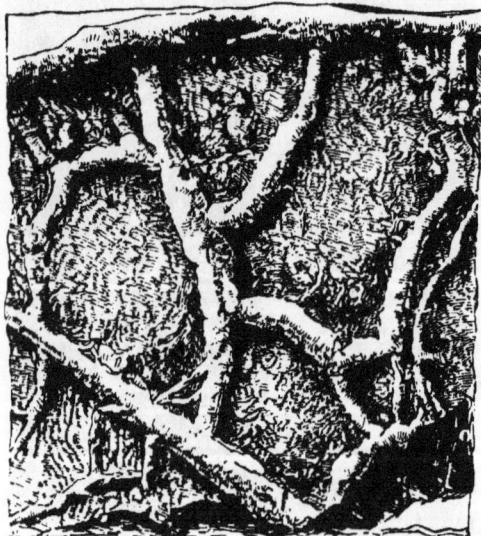

Fig. 11.—Cast of shrinkage-cracks (Carboniferous, Nova Scotia), illustrating pretended Algæ.

Slickensided bodies, resembling the fossil fruits described by Geinitz as *Gulielmites,* and the objects believed

by Fleming and Carruthers * to be casts of cavities filled
with fluid, abound in the shales of the Carboniferous and
Devonian. They are, no doubt, in most cases the results
of the pressure and consolidation of the clay around small
solid bodies, whether organic, fragmentary, or concretionary. They are, in short, local slickensides precisely
similar to those found so plentifully in the coal underclays, and which, as I have elsewhere † shown, resulted
from the internal giving way and slipping of the mass as
the roots of Stigmaria decayed within it. Most collectors
of fossil plants in the older formations must, I presume,
be familiar with appearances of this kind in connection
with small stems, petioles, fragments of wood, and car-
polites. I have in my collection petioles of ferns and
fruits of the genus *Trigonocarpum* partially slickensided
in this way, and which if wholly covered by this kind of
marking could scarcely have been recognised. I have
figured bodies of this kind in my report on the Devonian
and Upper Silurian plants of Canada, believing them,
owing to their carbonaceous covering, to be probably
slickensided fruits, though of uncertain nature. In every
case I think these bodies must have had a solid nucleus of
some sort, as the severe pressure implied in slickensiding
is quite incompatible with a mere "fluid-cavity," even
supposing this to have existed.

Prof. Marsh has well explained another phase of the
influence of hard bodies in producing partial slickensides,
in his paper on *Stylolites*, read before the American Association in 1867, and the application of the combined
forces of concretionary action and slickensiding to the
production of the cone-in-cone concretions, which occur
in the coal-formation and as low as the Primordial. I
have figured a very perfect and beautiful form of this

---

* "Journal of the Geological Society," June, 1871.

† *Ibid.*, vol. x., p. 14.

kind from the coal-formation of Nova Scotia, which is described in "Acadian Geology" * (Fig. 12).

I have referred to these facts here because they are relatively more important in that older period, which may be named the age of Algæ, and because their settlement now will enable us to dispense with discussions of this kind further on. The able memoirs of Nathorst and Williamson should be studied by those who desire further information.

But it may be asked, "Are there no real examples of fossil Algæ?" I believe there are many such, but the difficulty is to distinguish them. Confining ourselves to the older rocks, the following may be noted:

Fig. 12.—Cone-in-cone concretion (Carbon-iferous, Nova Scotia), illustrating pretended Algæ.

The genus *Buthotrephis* of Hall, which is characterised as having stems, sub-cylindric or compressed, with numerous branches, which are divaricating and sometimes leaf-like, contains some true Algæ. Hall's *B. gracilis*, from the Siluro-Cambrian, is one of these. Similar plants, referred to the same species, occur in the Clinton and Niagara formations, and a beautiful species, collected by Col. Grant, of Hamilton, and now in the McGill College collection, represents a broader and more frondose type of distinctly carbonaceous character. It may be described as follows:

Buthotrephis Grantii, S. N. (Fig. 13).—Stems and

---

* Appendix, p. 676, edition of 1878.

fronds smooth and slightly striate longitudinally, with curved and interrupted striæ. Stem thick, bifurcating, the divisions terminating in irregularly pinnate fronds, apparently truncate at the extremities. The quantity of carbonaceous matter present would indicate thick, though perhaps flattened, stems and dense fleshy fronds.

The species *Buthotrephis subnodosa* and *B. flexuosa*, from the Utica shale, are also certainly plants, though it is possible, if their structures and fruit were known, some of these might be referred to different genera. All of these plants have either carbonaceous matter or produce organic stains on the matrix.

FIG. 13.—*Buthotrephis Grantii*, a genuine Alga from the Silurian, Canada.

The organism with diverging wedge-shaped fronds, described by Hall as *Sphenothallus angustifolius*, is also a plant. Fine specimens, in the collection of the Geological Survey of Canada, show distinct evidence of the organic character of the wedge-shaped fronds. It is from the Utica shale, and elsewhere in the Siluro-Cambrian. It is just possible, as suggested by Hall, that this plant may be of higher rank than the Algæ.

The genus *Palæophycus* of Hall includes a great variety of uncertain objects, of which only a few are probably true Algæ. I have specimens of fragments similar to his *P. virgatus*, which show distinct carbonaceous films, and others from the Quebec group, which seem to be cylindrical tubes now flattened, and which have contained spindle-shaped sporangia of large size. Tortuous and curved flattened stems, or fronds, from the Upper Silurian limestone of Gaspé, also show organic matter.

Respecting the forms referred to *Licrophycus* by Billings, containing stems or semi-cylindrical markings springing from a common base, I have been in great doubt. I have not seen any specimens containing unequivocal organic matter, and am inclined to think that most of them, if not the whole, are casts of worm-burrows, with trails radiating from them.

Though I have confined myself in this notice to plants, or supposed plants, of the Lower Palæozoic, it may be well to mention the remarkable Cauda-Galli fucoids, referred by Hall to the genus *Spirophyton*, and which are characteristic of the oldest Erian beds. The specimens which I have seen from New York, from Gaspé, and from Brazil, leave no doubt in my mind that these were really marine plants, and that the form of a spiral frond, assigned to them by Hall, is perfectly correct. They must have been very abundant and very graceful plants of the early Erian, immediately after the close of the Silurian period.

We come now to notice certain organisms referred to Algæ, and which are either of animal origin, or are of higher grade than the sea-weeds. We have already discussed the questions relating to *Prototaxites*. *Drepano-phycus*, of Goeppert,[*] I suspect, is only a badly preserved branch or stem of the Erian land-plant known as *Arthro-stigma*. In like manner, *Haliserites Dechenianus*,[†] of Goeppert, is evidently the land-plant known as *Psilophy-ton*. *Sphærococcites dentatus* and *S. serra*—the *Fucoides dentatus* and *serra* of Brongniart, from Quebec—are graptolites of two species quite common there.[‡] *Dic-tyophyton* and *Uphantenia*, as described by Hall and the author, are now known to be sponges. They have become *Dictyospongiæ*. The curious and very ancient fossils referred by Forbes to the genus *Oldhamia* are perhaps still subject to doubt, but are usually regarded as Zoophytes, though it is quite possible they may be plants. Though I have not seen the specimens, I have no doubt whatever that the plants, or the greater part of them, from the Silurian of Bohemia, described by Stur as Algæ and Characeæ,[#] are really land-plants, some of them of the genus *Psilophyton*. I may say in this connection that specimens of flattened *Psilophyton* and *Arthrostig-ma*, in the Upper Silurian and Erian of Gaspé, would probably have been referred to Algæ, but for the fact that in some of them the axis of barred vessels is preserved.

It is not surprising that great difficulties have occurred in the determination of fossil Algæ. Enough, however, remains certain to prove that the old Cambrian and Silurian seas were tenanted with sea-weeds not very dissimilar from those of the present time. It is further probable that some of the graphitic, carbonaceous, and bituminous

---

[*] "Fossile Flora," 1852, p. 92, Table xli.

[†] *Ibid.*, p. 88, Table ii.

[‡] Brongniart, "Vegeteaux Fossiles," Plate vi., Figs. 7 to 12.

[#] "Proceedings of the Vienna Academy," 1881. *Hostinella*, of this author, is almost certainly *Psilophyton*, and his *Barrandiana* seems to include *Arthrostigma*, and perhaps leafy branches of *Berwynia*. These curious plants should be re-examined.

shales and limestones of the Silurian owe their carbonaceous matters to the decomposition of Algæ, though possibly some of it may have been derived from Graptolites
and other corneous Zoöphytes.   In any case, such micro-

FIG. 14.—Silurian vegetation restored.   *Protannularia, Berwynia, Nematophyton, Sphenophyllum, Arthrostigma, Psilophyton.*

scopic examinations of these shales as I have made, have
not produced any evidence of the existence of plants of
higher grade, while those of the Erian and Carboniferous
periods, similar to the naked eye, abound in such evidence.   It is also to be observed that, on the surfaces of

beds of sandstone in the Upper Cambrian, carbonaceous *débris*, which seems to be the remains of either aquatic or land plants, is locally not infrequent.

Referring to the land vegetation of the older rocks, it is difficult to picture its nature and appearance. We may imagine the shallow waters filled with aquatic or amphibious Rhizocarpean plants, vast meadows or brakes of the delicate *Psilophyton* and the starry *Protannularia* and some tall trees, perhaps looking like gigantic clubmosses, or possibly with broad, flabby leaves, mostly cellular in texture, and resembling Algæ transferred to the air. Imagination can, however, scarcely realise this strange and grotesque vegetation, which, though possibly copious and luxuriant, must have been simple and monotonous in aspect, and, though it must have produced spores and seeds and even fruits, these were probably all of the types seen in the modern acrogens and gymnosperms.

" In garments green, indistinct in the twilight,
They stand like Druids of old, with voices sad and prophetic."

Prophetic they truly were, as we shall find, of the more varied forests of succeeding times, and they may also help us to realise the aspect of that still older vegetation, which is fossilised in the Laurentian graphite; though it is not impossible that this last may have been of higher and more varied types, and that the Cambrian and Silurian may have been times of depression in the vegetable world, as they certainly were in the submergence of much of the land.

These primeval woods served at least to clothe the nakedness of the new-born land, and they may have sheltered and nourished forms of land-life still unknown to us, as we find as yet only a few insects and scorpions in the Silurian. They possibly also served to abstract from the atmosphere some portion of its superabundant carbonic acid harmful to animal life, and they stored up supplies of graphite, of petroleum, and of illuminating gas, useful to man at the present day. We may write of them and draw their forms with the carbon which they themselves supplied.

NOTE TO CHAPTER II.

EXAMINATION OF PROTOTAXITES (*Nematophyton*), BY PROF. PEN-HALLOW, OF McGILL UNIVERSITY.

Prof. Penhallow, having kindly consented to re-examine my specimens, has furnished me with elaborate notes of his facts and conclusions, of which the following is a summary, but which it is hoped will be published in full:

"1. *Concentric Layers.*—The inner face of each of these is composed of relatively large tubes, having diameters from 13·6 to 34·6 micro-millimetres. The outer face has tubes ranging from 13·8 to 27·6 mm. The average diameter in the lower surface approaches to 34, that in the outer to 13·8. There is, however, no abrupt termination to the surface of the layers, though in some specimens they separate easily, with shining surfaces.

"2. *Minute Structure.*—In longitudinal sections the principal part of the structure consists of longitudinal tubes of indeterminate length, and round in cross-section. They are approximately parallel, but in some cases may be seen to bend sinuously, and are not in direct contact. Finer myceloid tubes, 5·33 mm. in diameter, traverse the structure in all directions, and are believed to branch off from the larger tubes. In a small specimen supposed to be a branch or small stem, and in which the vertical tubes are somewhat distant from one another, this horizontal system is very largely developed; but is less manifest in the older stems. The tubes themselves show no structure. The ray-like openings in the substance of the tissue are evidently original parts of the structure, but not of the nature of medullary rays. They are radiating spaces running outward in an interrupted manner or so tortuously that they appear to be interrupted in their course from the centre towards the surface. They show tubes turning into them, branching into them, and approximately horizontal, but tortuous. On the external surface of some specimens these radial spaces are represented by minute pits irregularly or spirally arranged. The transverse swellings of the stem show no difference of structure, except that the tubes or cells may be a little more tortuous, and a transverse film of coaly matter extends from the outer coaly envelope inwardly. This may perhaps be caused by some accident of preservation. The outer coaly layer shows tubes similar to those of the stem.* The horizontal or oblique flexures of the large tubes seem to be mainly in the vicinity of the radial openings, and it is in entering these that they have been seen to branch."

The conclusions arrived at by Prof. Penhallow are as follows :

" 1. The plant was not truly exogenous, and the appearance of rings is independent of the causes which determine the layers of growth in exogenous plants.

" 2. The plant was possessed of no true bark. Whatever cortical layer was present was in all probability a modification of the general structure.†

" 3. An intimate relation exists between the large tubular cells and the myceloid filaments, the latter being a system of small branches from the former ; the branching being determined chiefly in certain special openings which simulate medullary rays.

" 4. The specimens examined exhibit no evidence of special decay, and the structure throughout is of a normal character.

" 5. The primary structure consists of large tubular cells without apparent terminations, and devoid of structural markings, with which is associated a secondary structure of myceloid filaments arising from the former.

" 6. The structure of *Nematophyton* as a whole is unique ; at least there is no plant of modern type with which it is comparable. Nevertheless, the loose character of the entire structure ; the interminable cells ; their interlacing ; and, finally, their branching into a secondary series of smaller filaments, point with considerable force to the true relationship of the stem as being with Algæ or other Thallophytes rather than with Gymnosperms. A more recent examination

---

* It is possible that these tubes may be merely part of the stem attached to the bark, which seems to me to indicate the same dense cellular structure seen in the bark of *Lepidodendra*, etc.

† On these points I would reserve the considerations : 1. That there must have been some relation between the mode of growth of these great stems and their concentric rings ; and, 2. That the evidence of a bark is as strong as in the case of any Palæozoic tree in which the bark is, as usual, carbonised.

of a laminated resinous substance found associated with the plant shows that it is wholly amorphous, and, as indicated by distinct lines of flow, that it must have been in a plastic state at a former period. The only evidence of structure was found in certain well-defined mycelia, which may have been derived from associated vegetable matter upon which they were growing, and over which the plastic matrix flowed."

I have only to add to this description that when we consider that *Nematophyton Logani* was a large tree, sometimes attaining a diameter of more than two feet, and a stature of at least twenty before branching; that it had great roots, and gave off large branches; that it was an aërial plant, probably flourishing in the same swampy flats with *Psilophyton, Arthrostigma,* and *Leptophleum ;* that the peculiar bodies known as *Pachytheca* were not unlikely its fruit—we have evidence that there were, in the early Palæozoic period, plants scarcely dreamt of by modern botany. Only when the appendages of these plants are more fully known can we hope to understand them. In the mean time, I may state that there were probably different species of these trees, indicated more particularly by the stems I have described as *Nematoxylon* and *Celluloxylon.** There were, I think, some indications that the plants described by Carruthers as *Berwynia,* may also be found to have been generically the same. The resinous matter mentioned by Prof. Penhallow is found in great abundance in the beds containing *Nematophyton,* and must, I think, have been an exudation from its bark.

---

* "Journal Geol. Society of London," 1863, 1881.

CHAPTER III.

In the last chapter we were occupied with the comparatively few and obscure remains of plants entombed
in the oldest geological formations. We now ascend to a
higher plane, that of the Erian or Devonian period, in
which, for the first time, we find varied and widely distributed forests.

The growth of knowledge with respect to this flora
has been somewhat rapid, and it may be interesting to
note its principal stages, as an encouragement to the hope
that we may yet learn something more satisfactory respecting the older floras we have just discussed.

In Goeppert's memoir on the flora of the Silurian,
Devonian, and Lower Carboniferous rocks, published in
1860,[*] he enumerates twenty species as Silurian, but these
are all admitted to be Algæ, and several of them are remains which may be fairly claimed by the zoölogists as
zoöphytes, or trails of worms and mollusks. In the Lower
Devonian he knows but six species, five of which are
Algæ, and the remaining one a *Sigillaria*, but this is of
very doubtful nature. In the Middle Devonian he gives
but one species, a land-plant of the genus *Lepidodendron*.
In the Upper Devonian the number rises to fifty-seven,
of which all but seven are terrestrial plants, representing

---

[*] Jena, 1860.

a large number of the genera occurring in the succeeding Carboniferous system.

Goeppert does not include in his enumeration the plants from the Devonian of Gaspé, described by the author in 1859,[*] having seen only an abstract of the paper at the time of writing his memoir, nor does he appear to have any knowledge of the plants of this age described by Lesquereux in Rogers's "Pennsylvania." These might have added ten or twelve species to his list, some of them probably from the Lower Devonian. It is further to be observed that a few additional species had also been recognised by Peach in the Old Red Sandstone of Scotland.

But from 1860 to the present time a rich harvest of specimens has been gathered from the Gaspé sandstones, from the shales of southern New Brunswick, from the sandstones of Perry in Maine, and from the wide-spread Erian areas of New York, Pennsylvania, and Ohio. Nearly all these specimens have passed through my hands, and I am now able to catalogue about a hundred species, representing more than thirty genera, and including all the great types of vascular Cryptogams, the Gymnosperms, and even one (still doubtful) Angiosperm. Many new forms have also been described from the Devonian of Scotland and of the Continent of Europe.

Before describing these plants in detail, we may refer to North America for illustration of the physical conditions of the time. In a physical point of view the northern hemisphere presented a great change in the Erian period. There were vast foldings of the crust of the earth, and great emissions of volcanic rock on both sides of the Atlantic. In North America, while at one time the whole interior area of the continent, as far north as

---

* "Journal of the Geological Society of London," also "Canadian Naturalist."

i, was occupied by a vast inland sea, studded
ids, the long Appalachian ridge had begun
g with the old Laurentian land, something
our present continent, and on the margins
hian belt there were wide, swampy flats and
ireas, which, under the mild climate that
characterised this period, were admirably
ish a luxuriant vegetation. Under this
lso, it would seem that new forms of plants
oduced in the far north, where the long
summer sunlight, along with great warmth,
aided in their introduction and early ex-
ience made their way to the southward, a
as Gray and others have shown, has also
er geological times.

ia of this Erian age consisted during the
the period of a more or less extensive belt
north with two long tongues descending
ong the Appalachian line in the east, the
egion west of the Rocky Mountains. On
des of these there were low lands covered
n, while on the inland side the great in-
i its verdant and wooded islands, realised,
ly with shallower water, the conditions of
hipelagoes of the Pacific.

sented conditions somewhat similar, having
nd middle portions of the period great sea
ilar patches of land, and later wide tracts
l in part enclosed water areas, swarming
d having an abundant vegetation on their
were the conditions of the Eifel and
iestones, and of the Old Red Sandstone of
the Kiltorcan beds of Ireland. In Europe
ierica, there were in the Erian age great
ieous rock. On both sides of the Atlantic
iewhat varied and changing conditions of

land and water, and a mild and equable climate, permitting the existence of a rich vegetation in high northern latitudes.   Of this latter fact a remarkable example is afforded by the beds holding plants of this age in Spitzbergen and Bear Island, in its vicinity.   Here there seem to be two series of plant-bearing strata, one with the vegetation of the Upper Erian, the other with that of the Lower Carboniferous, though both have been united by Heer under his so-called "Ursa Stage," in which he has grouped the characteristic plants of two distinct periods.   This has recently been fully established by the researches of Nathorst, though the author had already suggested it as the probable explanation of the strange union of species in the Ursa group of Heer.

In studying the vegetation of this remarkable period, we must take merely some of the more important forms as examples, since it would be impossible to notice all the species, and some of them may be better treated in the Carboniferous, where they have their headquarters. (Fig. 15.)

I may first refer to a family which seems to have culminated in the Erian age, and ever since to have occupied a less important place.   It is that of the curious aquatic plants known as Rhizocarps,* and referred to in the last chapter.

My attention was first directed to these organisms by the late Sir W. E. Logan in 1869.   He had obtained from the Upper Erian shale of Kettle Point, Lake Huron, specimens filled with minute circular discs, to which he referred, in his report of 1863, as "microscopic orbicular bodies."   Recognising them to be macrospores, or spore-cases, I introduced them into the report on the Erian

---

* Or, as they have recently been named by some botanists, "Heterosporous Filices," though they are certainly not ferns in any ordinary sense of that term.

flora, which I was then preparing, and which was published in 1871, under the name *Sporangites Huronensis*.

In 1871, having occasion to write a communication to the "American Journal of Science" on the question then

FIG. 15.—Vegetation of the Devonian period, restored. *Calamites, Psilophyton, Leptophleum, Lepidodendron, Cordaites, Sigillaria, Dadoxylon, Asterophyllites, Platyphyllum.*

raised as to the share of spores and spore-cases in the accumulation of coal, a question to be discussed in a subsequent chapter, these curious little bodies were again reviewed, and were described in substance as follows :

"The oldest bed of spore-cases known to me is that at Kettle Point, Lake Huron. It is a bed of brown bituminous shale, burning with much flame, and under a lens is seen to be studded with flattened disc-like bodies, scarcely more than a hundredth of an inch in diameter, which under the microscope are found to be spore-cases (or macrospores) slightly papillate externally (or more properly marked with dark pores), and sometimes showing a point of attachment on one side and a slit more or less elongated and gaping on the other. When slices of the rock are made, its substance is seen to be filled with these bodies, which, viewed as transparent objects, appear yellow like amber, and show little structure, except that the walls can be distinguished from the internal cavity, which may sometimes be seen to enclose patches of granular matter. In the shale containing them are also vast numbers of rounded, translucent granules, which may be escaped spores (microspores)." The bed containing these spores at Kettle Point was stated, in the reports of the "Geological Survey of Canada," to be twelve or fourteen feet in thickness, and besides these specimens it contained fossil plants referable to the species *Calamites inornatus* and *Lepidodendron primævum*, and I not unnaturally supposed that the Sporangites might be the fruit of the latter plant. I also noticed their resemblance to the spore-cases of *L. corrugatum* of the Lower Carboniferous (a Lepidodendron allied to *L. primævum*), and to those from Brazil described by Carruthers under the name *Flemingites*, as well as to those described by Huxley from certain English coals, and to those of the Tasmanite or white coal of Australia. The bed at Kettle Point is shown to be marine by its holding the sea-weed known as *Spirophyton*, and shells of *Lingula*.

The subject did not again come under my notice till

1882, when Prof. Orton, of Columbus, Ohio, sent me some specimens from the Erian shales of that State, which on comparison seemed undistinguishable from *Sporangites Huronensis*.* Prof. Orton read an interesting paper on these bodies, at the meeting of the American Association in Montreal, in which were some new and striking facts. One of these was the occurrence of such bodies throughout the black shales of Ohio, extending "from the Huron River, on the shore of Lake Erie, to the mouth of the Scioto, in the Ohio Valley, with an extent varying from ten to twenty miles in breadth," and estimated to be three hundred and fifty feet in thickness. I have since been informed by my friend Mr. Thomas, of Chicago, that its thickness, in some places at least, must be three times that amount. About the same time, Prof. Williams, of Cornell, and Prof. Clarke, of Northampton, announced similar discoveries in the State of New York, so that it would appear that beds of vast area and of great thickness are replete with these little vegetable discs, usually converted into a highly bituminous, amber-like substance, giving a more or less inflammable character to the containing rock.

Another fact insisted on by Prof. Orton was the absence of Lepidodendroid cones, and the occurrence of filamentous vegetable matter, to which the Sporangites seemed to be in some cases attached in groups. Prof. Orton also noticed the absence of the trigonal form, which belongs to the spores of many Lepidodendra, though this is not a constant character. In the discussion on Prof. Orton's paper, I admitted that the facts detailed by him shook my previous belief of the lycopodiaceous character

---

* These shales have been described, as to their chemical and geological relations, by Dr. T. Sterry Hunt, "American Journal of Science," 1863, and by Dr. Newberry, in the "Reports of the Geological Survey of Ohio," vol. i., 1863, and vol. iii., 1878.

of these bodies, and induced me to suspect, with Prof. Orton, that they might have belonged to some group of aquatic plants lower than the Lycopods.

Since the publication of my paper on Rhizocarps in the Palæozoic period above referred to, I have received two papers from Mr. Edward Wethered, F. G. S., in one of which he describes spores of plants found in the lower limestone shales of the Forest of Dean, and in the other discusses more generally the structure and origin of Carboniferous coal-beds.* In both papers he refers to the occurrence in these coals and shales of organisms essentially similar to the Erian spores.

In the "Bulletin of the Chicago Academy of Science," January, 1884, Dr. Johnson and Mr. Thomas, in their paper on the "Microscopic Organisms of the Boulder Clay of Chicago and Vicinity," notice *Sporangites Huronensis* as among these organisms, and have discovered them also in large numbers in the precipitate from Chicago city water-supply. They refer them to the decomposition of the Erian shales, of which boulders filled with these organisms are of frequent occurrence in the Chicago clays. The Sporangites and their accompaniments in the boulder clay are noticed in a paper by Dr. G. M. Dawson, in the "Bulletin of the Chicago Academy," June, 1885.

Prof. Clarke has also described, in the "American Journal of Science" for April, 1885, the forms already alluded to, and which he finds to consist of macrospores enclosed in sporocarps. He compares these with my *Sporangites Huronensis* and *Protosalvinia bilobata*, but I think it is likely that one of them at least is a distinct species.

I may add that in the "Geological Magazine" for 1875, Mr. Newton, F. G. S., of the Geological Survey of

---

* "Cotteswold Naturalists' Field Club," 1884; "Journal of the Royal Microscopical Society," 1885.

England, published a description of the Tasmanite and Australian white coal, in which he shows that the organisms in these deposits are similar to my *Sporangites Huronensis*, and to the macrospores previously described by Prof. Huxley, from the Better-bed coal. Mr. Newton does not seem to have been aware of my previous description of *Sporangites*, and proposes the name *Tasmanites punctatus* for the Australian form.

Here we have the remarkable fact that the waste macrospores, or larger spores of a species of Cryptogamous plant, occur dispersed in countless millions of tons through the shales of the Erian in Canada and the United States.

No certain clue seemed to be afforded by all these observations as to the precise affinities of these widely distributed bodies; but this was furnished shortly after from an unexpected quarter. In March, 1883, Mr. Orville Derby, of the Geological Survey of Brazil, sent me specimens found in the Erian of that country, which seemed to throw a new light on the whole subject. These I described and pointed out their connection with *Sporangites* at the meeting of the American Association at Minneapolis, in 1883, and subsequently published my notes respecting them in its proceedings, and in the "Canadian Record of Science."

Mr. Derby's specimens contained the curious spiral sea-weed known as *Spirophyton*, and also minute rounded Sporangites like those obtained in the Erian of Ohio, and of which specimens had been sent to me some years before by the late Prof. Hartt. But they differed in showing the remarkable fact that these rounded bodies are enclosed in considerable numbers in spherical and oval sacs, the walls of which are composed of a tissue of hexagonal cells, and which resemble in every respect the involucres or spore-sacs of the little group of modern acrogens known as Rhizocarps, and living in shallow

water.   More especially they resemble the sporocarps of
the genus *Salvinia*.   This fact opened up an entirely
new field of investigation, and I at once proceeded to
compare the specimens with the fructification of modern
Rhizocarps, and found that substantially these multitu-
dinous spores embedded in the Erie shales may be regarded as perfectly analogous to the larger spores of the
modern *Salvinia natans* of Europe, as may be seen by
the representation of them in Fig. 16.

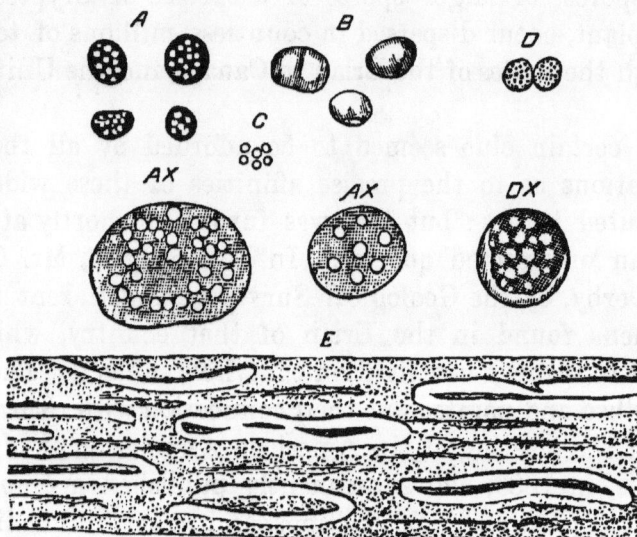

Fig. 16.—*Sporangites* (*Protosalvinia*).   A, *Sporangites Braziliensis*, natural
size.   AX, Same, magnified.   B, *Sp. biloba*, natural size.   C, Detached
macrospores.   D, Spore-cases of *Salvinia natans*.   DX, Same, magnified.
E, Shale with sporangites, vertical section, highly magnified.

The typical macrospores from the Erian shales are
perfectly circular in outline, and in the flattened state appear as discs with rounded edges, their ordinary diameter
being from one seventy-fifth to one one hundredth of an
inch, though they vary considerably in size.   This, however, I do not regard as an essential character.   The
edges, as seen in profile, are smooth, but the flat surface
often presents minute dark spots, which at first I mis-

took for papillæ, but now agree with Mr. Thomas in recognising them as minute pores traversing the wall of the
disc, and similar to those which Mr. Newton has described
in Tasmanite, and which Mr. Wethered has also recognised in the similar spores of the Forest of Dean shales.
The walls also sometimes show faint indications of concentric lamination, as if they had been thickened by successive deposits.

As seen by transmitted light, and either in front or in
profile, the discs are of a rich amber colour, translucent
and structureless, except the pores above referred to.
The walls are somewhat thick, or from one-tenth to one-
twentieth the diameter of the disc in thickness.   They
never exhibit the triradiate marking seen in spores of Lycopods, nor any definite point of attachment, though
they sometimes show a minute elongated spot which may
be of this nature, and they are occasionally seen to have
opened by slits on the edge or front, where there would
seem to have been a natural line of dehiscence.   The interior is usually quite vacant or structureless, but in some
cases there are curved internal markings which may indicate a shrunken lining membrane, or the remains of a
prothallus or embryo.   Occasionally a fine granular substance appears in the interior, possibly remains of microspores.

The discs are usually detached and destitute of any
envelope, but fragments of flocculent cellular matter are
associated with them, and in one specimen from the corniferous limestone of Ohio, in Mr. Thomas's collection, I
have found a group of eight or more discs partly enclosed
in a cellular sac-like membrane of similar character to
that enclosing the Brazilian specimens already referred to.

The characters of all the specimens are essentially
similar, and there is a remarkable absence of other organisms in the shale.   In one instance only, I have observed
a somewhat smaller round body with a dark centre or

nucleus, and a wide translucent margin, marked by a slight granulation. Even this, however, may indicate nothing more than a different state of preservation.

It is proper to observe here that the wall or enclosing sac of these macrospores must have been of very dense consistency, and now appears as a highly bituminous substance, in this agreeing with that of the spores of Lycopods, and, like them, having been when recent of a highly carbonaceous and hydrogenous quality, very combustible and readily admitting of change into bituminous matter. In the paper already referred to, on spore-cases in coals, I have noticed that the relative composition of lycopodium and cellulose is as follows :

Cellulose, $C_{24}H_{20}O_{20}$.

Lycopodium, $C_{42}H_{19}NO_{5}$.

Thus, such spores are admirably suited for the production of highly carbonaceous or bituminous coals, etc.

Nothing is more remarkable in connection with these bodies than their uniformity of structure and form over so great areas and throughout so great thickness of rock, and the absence of any other kind of spore-case. This is more especially noteworthy in contrast with the coarse coals and bituminous shales of the Carboniferous, which usually contain a great variety of spores and sporangia, indicating the presence of many species of acrogenous plants, while the Erian shales, on the contrary, indicate the almost exclusive predominance of one form. This contrast is well seen in the Bedford shales overlying these beds, and I believe Lower Carboniferous.* Specimens of these have been kindly communicated to me by Prof. Orton, and have been prepared by Mr. Thomas. In these we see the familiar Carboniferous spores with triradiate markings called *Triletes* by Reinsch, and which are similar to those of Lycopodiaceous plants. Still more abun-

---

* According to Newberry, lower part of Waverly group.

dant are those spinous and hooked spores or sporangia, to which the names *Sporocarpon*, *Zygosporites*, and *Traquaria* have been given, and some of which Williamson has shown to be spores of Lycopodiaceous plants.*

The true "Sporangites," on the contrary, are round and smooth, with thick bituminous walls, which are punctured with minute transverse pores. In these respects, as already stated, they closely resemble the bodies found in the Australian white coal and Tasmanite. The precise geological age of this last material is not known with certainty, but it is believed to be Palæozoic.

With reference to the mode of occurrence of these bodies, we may note first their great abundance and wide distribution. The horizontal range of the bed at Kettle Point is not certainly known, but it is merely a northern outlier of the great belt of Erian shales referred to by Prof. Orton, and which extends, with a breadth of ten to twenty miles, and of great thickness, across the State of Ohio, for nearly two hundred miles. This Ohio black shale, which lies at the top of the Erian or the base of the Carboniferous, though probably mainly of Erian age, appears to abound throughout in these organisms, and in some beds to be replete with them. In like manner, in Brazil, according to Mr. Derby, these organisms are distributed over a wide area and throughout a great thickness of shale holding *Spirophyton*, and apparently belonging to the Upper Erian. The recurrence of similar forms in the Tasmanite and white coal of Tasmania and Australia is another important fact of distribution. To this

---

* *Traquaria* is to be distinguished from the calcareous bodies found in the corniferous limestone of Kelly's Island, which I have described in the "Canadian Naturalist" as *Saccamina Eriana*, and believe to be Foraminiferal tests. They have since been described by Ulrich under a different name (*Mœllerina*: contribution to "American Palæontology," 1886). See Dr. Williamson's papers in "Transactions of Royal Society of London."

we may add the appearance of these macrospores in coals and shales of the Carboniferous period, though there in association with other forms.

It is also to be observed that the Erian shales, and the Forest of Dean beds described by Wethered, are marine, as shown by their contained fossils; and, though I have no certain information as to the Tasmanite and Australian white coal, they would seem, from the description of Milligan, to occur in distinctly aqueous, possibly estuarine, deposits. Wethered has shown that the discs described by Huxley and Newton in the Better-bed coal occur in the earthy or fragmentary layers, as distinguished from the pure coal. Those occurring in cannel coal are in the same case, so that the general mode of occurrence implies water-driftage, since, in the case of bodies so large and dense, wind-driftage to great distances would be impossible.

These facts, taken in connection with the differences between these macrospores and those of any known land-plant of the Palæozoic, would lead to the inference that they belonged to aquatic plants, and these vastly abundant in the waters of the Erian and Carboniferous periods.

It is still further to be observed that they are not, in the Erian beds, accompanied with any remains of woody or scalariform tissues, such as might be expected in connection with the *débris* of terrestrial acrogens, and that, on the other hand, we find them enclosed in cellular sporocarps, though in the majority of cases these have been removed by dehiscence or decay.

These considerations, I think, all point to the probability which I have suggested in my papers on this subject referred to above, that we have in these objects the organs of fructification of plants belonging to the order *Rhizocarpeæ*, or akin to it. The comparisons which I have instituted with the sporocarps and macrospores of these plants confirm this suggestion. Of the modern species which I have had an opportunity to examine, *Salvinia natans* of Europe perhaps presents the closest resemblance. In this plant groups of round cellular sporocarps appear at the bases of the floating fronds. They are about a line in diameter when mature, and are of two kinds, one containing macrospores, the other microspores or antheridia. The first, when mature, hold a number of closely packed globular or oval sporangia of loose cellular tissue, attached to a central placenta. Each of these sporangia contains a single macrospore, perfectly globular and smooth, with a dense outer membrane (exhibiting traces of lamination, and showing within an irregularly vacuolated or cellular structure, probably a prothallus). I cannot detect in it the peculiar pores which appear in the fossil specimens. Each macrospore is about one-seventieth of an inch in diameter when mature. The sporocarps of the microspores contain a vastly greater number of minute sporangia, about one two-hundredths of an inch in diameter. These contain disc-like antheridia, or microspores of very minute size.

The discs from Kettle Point and from the Ohio black shale, and from the shale boulders of the Chicago clays, are similar to the macrospores of *Salvinia*, except that they have a thicker wall and are a little less in diameter, being about one-eightieth of an inch. The Brazilian sporocarps are considerably larger than those of the modern *Salvinia*, and the macrospores approach in size to those of the modern species, being one seventy-fifth of an inch in diameter. They also seem, like the modern species, to have thinner walls than those from Canada, Ohio, and Chicago. No distinct indication has been observed in the fossil species of the inner Sporangium of *Salvinia*. Possibly it was altogether absent, but more probably it is not preserved as a distinct structure.

With reference to the microspores of *Salvinia*, it is to be observed that the sporocarps, and the contained spores

or antheridia, are very delicate and destitute of the dense
outer wall of the macrospores.   Hence such parts are
little likely to have been preserved in a fossil state ; and
in the Erian shales, if present, they probably appear
merely as flocculent carbonaceous matter not distinctly
marked, or as minute granules not well defined, of which
there are great quantities in some of the shales.

The vegetation appertaining to the Sporangites has
not been distinctly recognised.   I have, however, found
in one of the Brazilian specimens two sporocarps attached
to what seems a fragment of a cellular frond, and numerous specimens of the supposed Algæ, named *Spirophyton*,
are found in the shales, but there is no evidence of any
connection of this plant with the *Protosalvinia*.

Modern Rhizocarps present considerable differences as
to their vegetative parts.   Some, like *Pilularia*, have
simple linear leaves ; others, like *Marsilea*, have leaves in
whorls, and cuneate in form ; while others, like *Azolla*
and *Salvinia*, have frondose leaves, more or less pinnate
in their arrangement.   If we inquire as to fossils representing these forms of vegetation, we shall find that some
of the plants to be noticed in the immediate sequel may
have been nearly allied to the Rhizocarps.   In the mean
time I may state that I have proposed the generic name
Protosalvinia for these curious macrospores and their
coverings, and have described in the paper in the "Bulletin of the Chicago Academy of Sciences," already
quoted, five species which may be referred to this genus.

These facts lead to inquiries as to the origin of the
bituminous matter which naturally escapes from the
rocks of the earth as petroleum and inflammable gas, or
which may be obtained from certain shales in these forms
by distillation.   These products are compounds of carbon
and hydrogen, and may be procured from recent vegetable
substances by destructive distillation.   Some vegetable
matters, also, are much richer in carbon and hydrogen

than others, and it is a remarkable fact that the spores of
certain cryptogamous plants are of this kind, as we see in
the inflammable character of the dry spores of Lycopodium ; and we know that the slow putrefaction of such
material underground effects chemical changes by which
bituminous matter can be produced.   There is, therefore, nothing unreasonable in the supposition advanced
by Prof. Orton, that the spores so abundantly contained
in the Ohio black shales are important or principal sources
of the bituminous matter which they contain.   Microscopic sections of this shale show that much of its material consists of the rich bituminous matter of these spores
(Fig. 16).   At the same time, while we may trace the
bitumen of these shales, and of some beds of coal, to this
cause, we must bear in mind that there are other kinds of
bituminous rocks which show no such structures, and may
have derived their combustible material from other kinds
of vegetable matter, whether of marine or of land plants.
We shall better understand this when we have considered
the origin of coal.

The macrospores above referred to may have belonged
to humble aquatic plants mantling the surfaces of water
or growing up from the bottom, and presenting little
aërial vegetation.   But there are other Erian plants, as
already mentioned, which, while of higher structure, may
be of Rhizocarpean affinities.

One of these is the beautiful plant with whorls of
wedge-shaped leaves, to which the name *Sphenophyllum*
(see Fig. 20) has been given.   Plants referred to this
genus have been described by Lesquereux from the upper
part of the Siluro-Cambrian,[*] and a beautiful little species occurs in the Erian shales of St. John, New Brunswick.[†]   The genus is also continued, and is still more

---

[*] " American Journal of Science."
[†] Dawson, " Report on Devonian Plants," 1870.

abundant, in the Carboniferous.   Many years ago I observed, in a beautiful specimen collected by Sir W. E.
Logan, in New Brunswick, that the stem of this plant
had an axis of reticulated and scalariform vessels, and an
outer bark.* Renault and Williamson have more recently
obtained more perfect specimens, and the former has
figured a remarkably complex triangular axis, containing
punctate and barred vessels, and larger punctate vessels
filling in its angles.   Outside of this there is a cellular
inner bark, and this is surrounded by a thick fibrous envelope.   That a structure so complex should belong to
a plant so humble in its affinities is one of the strange
anomalies presented by the old world, and of which we
shall find many similar instances.   The fruit of *Sphenophyllum* was borne in spikes, with little whorls of bracts
or rudimentary leaves bearing round sporocarps.

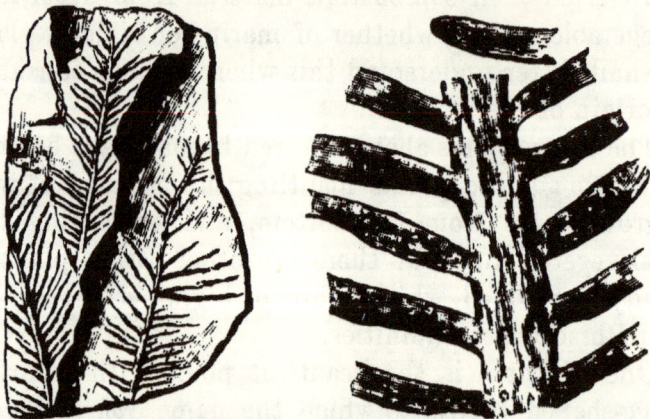

Fig. 17.—*Ptilophyton plumosum* (Lower Carboniferous, Nova Scotia).
Natural size and magnified.

A second type of plant, which may have been Rhizocarpean in its affinities, is that to which I have given the
name *Ptilophyton.*† It consists of beautiful feathery

---

* "Journal of the Geological Society," 1865.
† *Plumalina* of Hall.

fronds, apparently bearing on parts of the main stem or petiole small rounded sporocarps. They are found abundantly in the Middle Erian of the State of New York, and also occur in Scotland, while one species appears to occur in Nova Scotia, as high as the Lower Carboniferous (Figs. 17, 18).

These organisms have been variously referred to Lycopods, to Algæ, or to Zoöphytes, but an extended compari-

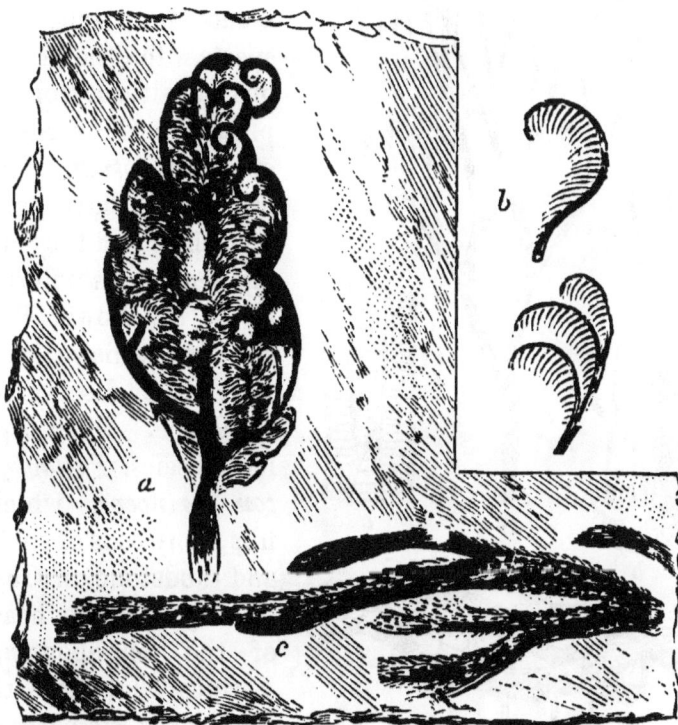

FIG. 18.—*Ptilophyton Thomsoni* (Scotland). *a*, Impression of plant in vernation. *b*, Branches conjecturally restored. *c*, Branches of *Lycopodites Milleri*, on same slab.

son of American and Scottish specimens has led me to the belief that they were aquatic plants, more likely to have been allied to Rhizocarps than to any other group. Some evidence of this will be given in a note appended to this chapter.

Fig. 19. — *Psilophyton princeps*, restored (Lower Erian, Gaspé). *a*, Fruit, natural size. *b*, Stem, natural size. *c*, Scalariform tissue of the axis, highly magnified. In the restoration, one side is represented in vernation and the other in fruit.

Another genus, which I have named *Psilophyton* * (Figs. 19, 21), may be regarded as a connecting link between the Rhizocarps and the Lycopods. It is so named from its resemblance, in some respects, to the curious parasitic Lycopods placed in the modern genus *Psilotum*. Several species have been described, and they are eminently characteristic of the Lower Erian, in which they were first discovered in Gaspé. The typical species, *Psilophyton princeps*, which fills many beds of shale and sandstone in Gaspé Bay and the head of the neighbouring Bay des Chaleurs with its slender stems and creeping, cord-like rhizomes, may be thus described :

Stems branching

* "Journal of the Geological Society," vols. xv., xviii., and xix., "Report on Devonian Plants of Canada," 1871.

dichotomously, and covered with interrupted ridges. Leaves rudimentary, or short, rigid, and pointed; in barren stems, numerous and spirally arranged; in fertile stems and branchlets, sparsely scattered or absent; in decorticated specimens, represented by minute punctate scars. Young branches circinate; rhizomata cylindrical, covered with hairs or ramenta, and having circular areoles irregularly disposed, giving origin to slender cylindrical rootlets. Internal structure—an axis of scalariform vessels, surrounded by a cylinder of parenchymatous cells, and by an outer cylinder of elongated woody cells. Fructification consisting of naked oval spore-cases, borne usually in pairs on slender, curved pedicels, either lateral or terminal.

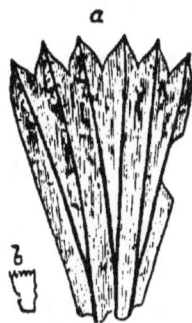

FIG. 20. — *Sphenophyllum antiquum* (Erian, New Brunswick). See pp. 61, 67.

This species was fully described by me in the papers referred to above, from specimens obtained from the rich exposures at Gaspé Bay, and which enabled me to illustrate its parts more fully, perhaps, than those of any other species of so great antiquity. In the specimens I had obtained I was able to recognise the forms of the rhizomata, stems, branches, and rudimentary leaves, and also the internal structure of the stems and rhizomata, and to illustrate the remarkable resemblance of the forms and structures to those of the modern *Psilotum*. The fructification was, however, altogether peculiar, consisting of narrowly ovate sporangia, borne usually in pairs, on curved and apparently rigid petioles. Under the microscope these sporangia show indications of cellular structure, and appear to have been membranous in character. In some specimens dehiscence appears to have taken place by a slit in one side, and, clay having entered into the interior, both walls of the spore-case can be seen. In other instances, being flattened, they might be mis-

taken for scales.   No spores could be observed in any of
the specimens, though in some the surface was marked
by slight, rounded prominences, possibly the impressions
of the spores within.   This peculiar and very simple style

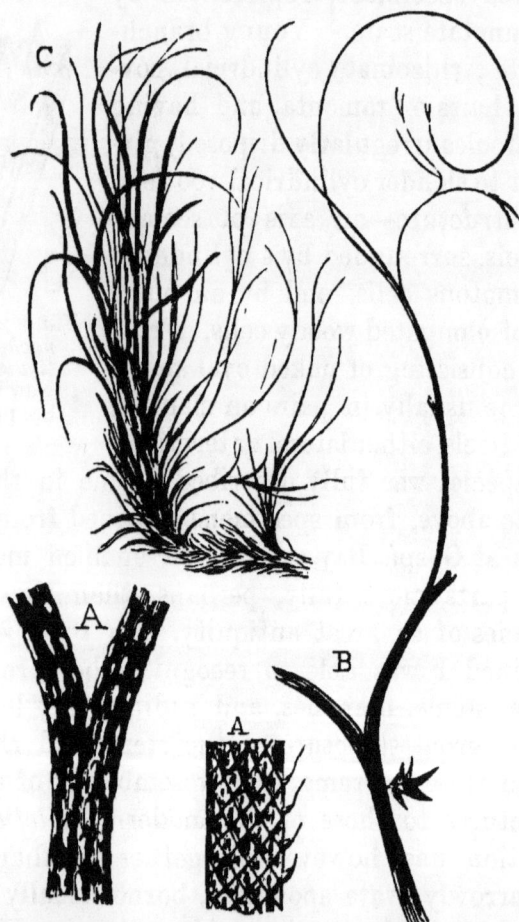

FIG. 21.—*Lepidodendron* and *Psilophyton* (Erian, New Brunswick).
A, *Lepidodendron Gaspianum*. B, C, *Psilophyton elegans*.

of spore-case is also characteristic of other species, and
gives to *Psilophyton* a very distinct generic character.
These naked spore-cases may be compared to those of
such lycopodiaceous plants as *Psilotum*, in which the

scales are rudimentary. They also bear some resemblance, though on a much larger scale, to the spore-cases of some Erian ferns (*Archæopteris*), to be mentioned in the sequel. On the whole, however, they seem most nearly related to the sporocarps of the Rhizocarpeæ.

Arthrostigma, which is found in the same beds with *Psilophyton,* was a plant of more robust growth, with better-developed, narrow, and pointed leaves, borne in a verticillate or spiral manner, and bearing at the ends of its branches spikes of naked sporocarps, apparently similar to those of *Psilophyton* but more rounded in form. The two genera must have been nearly related, and the slender branchlets of *Arthrostigma* are, unless well preserved, scarcely distinguishable from the stems of *Psilophyton.**

If, now, we compare the vegetation of these and similar ancient plants with that of modern Rhizocarps, we shall find that the latter still present, though in a depauperated and diminished form, some of the characteristics of their predecessors. Some, like *Pilularia,* have simple linear leaves ; others, like *Marsilea,* have leaves in verticils and cuneate in form ; while others, like *Azolla* and *Salvinia,* have frondose leaves, more or less pinnate in their arrangement. The first type presents little that is characteristic, but there are in the Erian sandstones and shales great quantities of filamentous and linear objects which it has been impossible to refer to any genus, and which might have belonged to plants of the type of *Pilularia.* It is quite possible, also, that such plants as *Psilophyton glabrum* and *Cordaites angustifolia,* of which the fructification is quite unknown, may have been allied to Rhizocarps. With regard to the verticillate type, we are at once reminded of *Sphenophyllum* (Fig. 20), which

---

* Reports of the author on " Devonian Plants," " Geological Survey of Canada," which see for details as to Erian Flora of northeastern America.

many palæo-botanists have referred to the *Marsiliaceæ*, though, like other Palæozoic Acrogens, it presents complexities not seen in its modern representatives. *S. primævum* of Lesquereux is found in the Hudson River group, and my *S. antiquum* in the Middle Erian. Besides these, there are in the Silurian and Erian beds plants with verticillate leaves which have been placed with the Annulariæ, but which may have differed from them in fructification. *Annularia laxa,* of the Erian, and *Protannularia Harknessii,* of the Siluro-Cambrian, may be given as examples, and must have been aquatic plants, probably allied to Rhizocarps. It is deserving of notice, also, that the two best-known species of *Psilophyton* (*P. princeps* and *P. robustius*), while allied to Lycopods by the structure of the stem and such rudimentary foliage as they possess, are also allied, by the form of their fructification, to the Rhizocarps, and not to ferns, as some palæo-botanists have incorrectly supposed. A similar remark applies to *Arthrostigma ;* and the beautiful pinnately leaved *Ptilophyton* may be taken to represent that type of foliage as seen in modern Rhizocarps, while the allied forms of the Carboniferous which Lesquereux has named *Trochophyllum,* seem to have had sporocarps attached to the stem in the manner of *Azolla.*

The whole of this evidence, I think, goes to show that in the Erian period there were vast quantities of aquatic plants, allied to the modern Rhizocarps, and that the so-called *Sporangites* referred to in this paper were probably the drifted sporocarps and macrospores of some of these plants, or of plants allied to them in structure and habit, of which the vegetative organs have perished. I have shown that in the Erian period there were vast swampy flats covered with *Psilophyton,* and in similar submerged tracts near to the sea the *Protosalvinia* may have filled the waters and have given off the vast multitudes of macrospores which, drifted by currents, have settled in the

mud of the black shales.　We have thus a remarkable
example of a group of plants reduced in modern times to
a few insignificant forms, but which played a great *rôle* in
the ancient Palæozoic world.

Leaving the Rhizocarps, we may now turn to certain
other families of Erian plants.　The first to attract our
attention in this age would naturally be the Lycopods,
the club-mosses or ground-pines, which in Canada and
the Eastern States carpet the ground in many parts of
our woods, and are so available for the winter decoration
of our houses and public buildings.　If we fancy one of
these humble but graceful plants enlarged to the dimensions of a tree, we shall have an idea of a *Lepidodendron*,
or of any of its allies (Figs. 15, 21).　These large lycopodiaceous trees, which in different specific and generic
forms were probably dominant in the Erian woods, resembled in general those of modern times in their fruit
and foliage, except that their cones were large, and probably in most cases with two kinds of spores, and their
leaves were also often very long, thus bearing a due proportion to the trees which they clothed.　Their thick
stems required, however, more strength than is necessary
in their diminutive successors, and to meet this want
some remarkable structures were introduced similar to
those now found only in the stems of plants of higher
rank.　The cells and vessels of all plants consist of thin
walls of woody matter, enclosing the sap and other contents of these sacs and tubes, and when strength is required it is obtained by lining their interior with successive coats of the hardest form of woody matter, usually
known as lignin.　But while the walls remain thin, they
afford free passage to the sap to nourish every part.　If
thickened all over, they would become impervious to sap,
and therefore unsuited to one of their most important
functions.　These two ends of strength and permeability
are secured by partial linings of lignin, leaving portions of

the original wall uncovered.   But this may be done in a great variety of ways.

The most ancient of these contrivances, and one still continued in the world of plants, is that of the barred or scalariform vessel.   This may be either square or hexagonal, so as to admit of being packed without leaving vacancies.   It is strengthened by a thick bar of ligneous matter up each angle, and these are connected by cross-bars so as to form a framework resembling several ladders fastened together.   Hence the name *scalariform*, or ladder-like.   Now, in a modern Lycopod there is a central axis of such barred vessels associated with simpler fibres or elongated cells.   Even in *Sphenophyllum* and *Psilophyton*, already referred to as allied to Rhizocarps,[*] there is such a central axis, and in the former rigidity is given to this by the vascular and woody elements being arranged in the form of a three-sided prism or three-rayed star.   But such arrangements would not suffice for a tree, and hence in the arboreal Lycopods of the Erian age a more complex structure is introduced.   The barred vessels were expanded in the first instance into a hollow cylinder filled in with pith or cellular tissue, and the outer rind was strengthened with greatly thickened cells. But even this was not sufficient, and in the older stems wedge-shaped bundles of barred tissue were run out from the interior, forming an external woody cylinder, and inside of the rind were placed bundles of tough bast fibres. Thus, a stem was constructed having pith, wood, and bark, and capable of additions to the exterior of the woody wedges by a true exogenous growth.   The plan is, in short, the same with that of the stems of the exogenous trees of modern times, except that the tissues employed are less complicated.   The structures of these remarkable

---

[*] First noticed by the author, "Journal of Geological Society," 1865; but more completely by Renault, "Comptes Rendus," 1870.

trees, and the manner in which they anticipate those of the true exogens of modern times, have been admirably illustrated by Dr. Williamson, of Manchester.   His papers, it is true, refer to these plants as existing in the Carboniferous age, but there is every reason to believe that they were of the same character in the Erian.   The plan is the same with that now seen in the stems of exogenous phænogams, and which has long ceased to be used in those of the Lycopods.   In this way, however, large and graceful lycopodiaceous trees were constructed in the Erian period, and constituted the staple of its forests.

The roots of these trees were equally remarkable with their stems, and so dissimilar to any now existing that botanists were long disposed to regard them as independent plants rather than roots.   They were similar in general structure to the stems to which they belonged, but are remarkable for branching in a very regular manner by bifurcation like the stems above, and for the fact that their long, cylindrical rootlets were arranged in a spiral manner and distinctly articulated to the root after the manner of leaves rather than of rootlets, and fitting them for growing in homogeneous mud or vegetable muck.   They are the so-called *Stigmaria* roots, which, though found in the Erian and belonging to its lycopodiaceous plants, attained to far greater importance in the Carboniferous period, where we shall meet with them again.

There were different types of lycopodiaceous plants in the Erian.   In addition to humble Lycopods like those of our modern woods and great Lepidodendra, which were exaggerated Lycopods, there were thick-stemmed and less graceful species with broad rhombic scars (*Leptophleum*), and others with the leaf-scars in vertical rows (*Sigillaria*), and others, again, with rounded leaf-scars, looking like the marks on Stigmaria, and belonging to the genus *Cyclostigma*.   Thus some variety was given to the arboreal club-mosses of these early forests.   (See Fig. 15.)

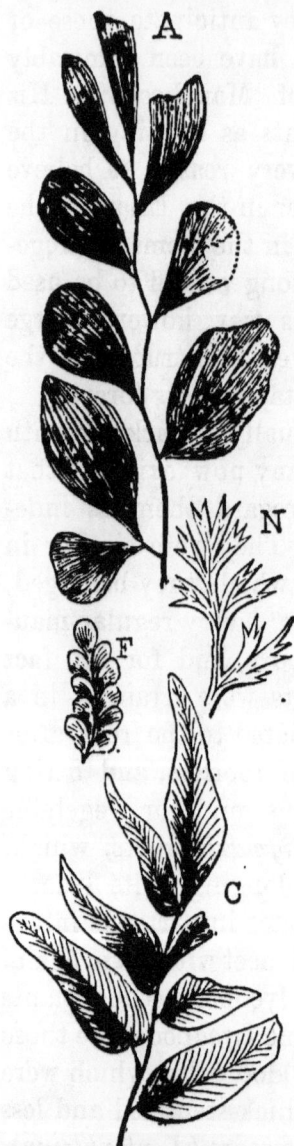

FIG. 22.—Erian ferns (New Brunswick). A, *Aneimites obtusa*. O, *Neuropteris polymorpha*. F, *Sphenopteris pilosa*. N, *Hymenophyllites subfurcatus*.

Another group of plants which attained to great development in the Erian age is that of the Ferns or Brackens. The oldest of these yet known are found in the Middle Erian. The *Eopteris* of Saporta, from the Silurian, at one time supposed to carry this type much further back, has unfortunately been found to be a mere imitative form, consisting of films of pyrites of leaf-like shapes, and produced by crystallisation. In the Middle Erian, however, more especially in North America, many species have been found (Figs. 22 to 24).* I have myself recorded more than thirty species from the Middle Erian of Canada, and these belong to several of the genera found in the Carboniferous, though some are peculiar to the Erian. Of the latter, the best known are perhaps those of the genus *Archæopteris* (Fig. 24), so abundant in the plant-beds of Kiltorcan in Ireland, as well as in North America. In this genus the fronds are large and luxuriant, with broad obovate pinnules decurrent on the leaf-stalk, and with simple sac-like spore-cases borne on modified pinnæ. Another very beautiful fern found

---

* For descriptions of these ferns, see reports cited above.

FIG. 28.—Erian ferns (New Brunswick). B, *Cyclopteris valida*, and pinnule enlarged. D, *Sphenopteris marginata*, and portion enlarged. E, *Sphenopteris Hartii*. G, *Hymenophyllites curtilobus*. H, *Hymeno-phyllites Gersdorffii*, and portion enlarged. I, *Alethopteris discrepans*. K, *Pecopteris serrulata*. L, *Pecopteris preciosa*. M, *Alethopteris Perleyi*.

with *Archæopteris* is that which I have named *Platyphyllum*, and which grew on a creeping stem or parasitically on stems of other plants, and had marginal fructification.*

FIG. 24.—*Archæopteris Jacksoni*, Dawson (Maine).   An Upper Erian fern.   *a*, *b*, Pinnules showing venation.

---

* "Reports on Fossil Plants of the Devonian and Upper Silurian of Canada," 1871, &c.

Another very remarkable fern, which some botanists have supposed may belong to a higher group than the ferns, is Megalopteris (Fig. 26).

Some of the Erian ferns attained to the dimensions of tree-ferns. Large stems of these, which must have floated out far from land, have been found by Newberry in the marine limestone of Ohio (*Caulopteris antiqua* and *C. peregrina*, Newberry),* and Prof. Hall has found in the

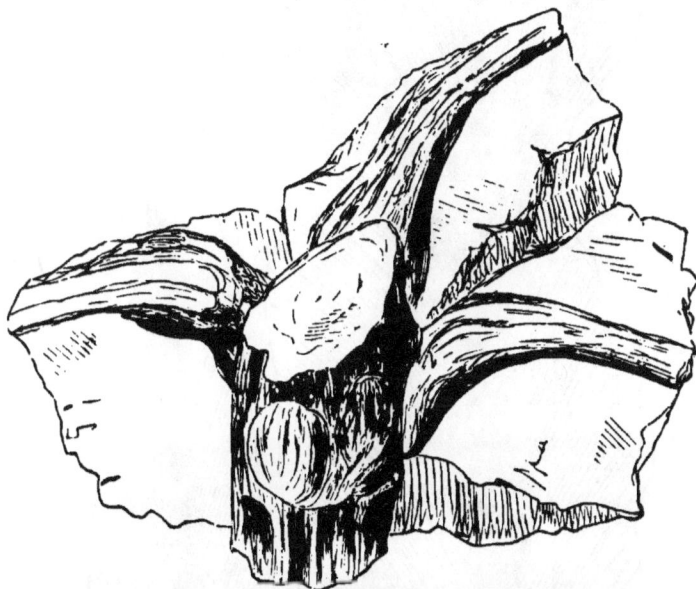

FIG. 25.—An Erian tree-fern. *Caulopteris Lockwoodi*, Dawson, reduced. (From a specimen from Gilboa, New York.)

Upper Devonian of Gilboa, New York, the remains of a forest of tree-ferns standing *in situ* with their great masses of aërial roots attached to the soil in which they grew (*Caulopteris Lockwoodi*, Dn.).†

These aërial roots introduce us to a new contrivance for strengthening the stems of plants by sending out into the soil multitudes of cord-like cylindrical roots from

---

* "Journal of the Geological Society," 1871. † *Ibid.*

various heights on the stem, and which form a series of
stays like the cordage of a ship.   This method of support

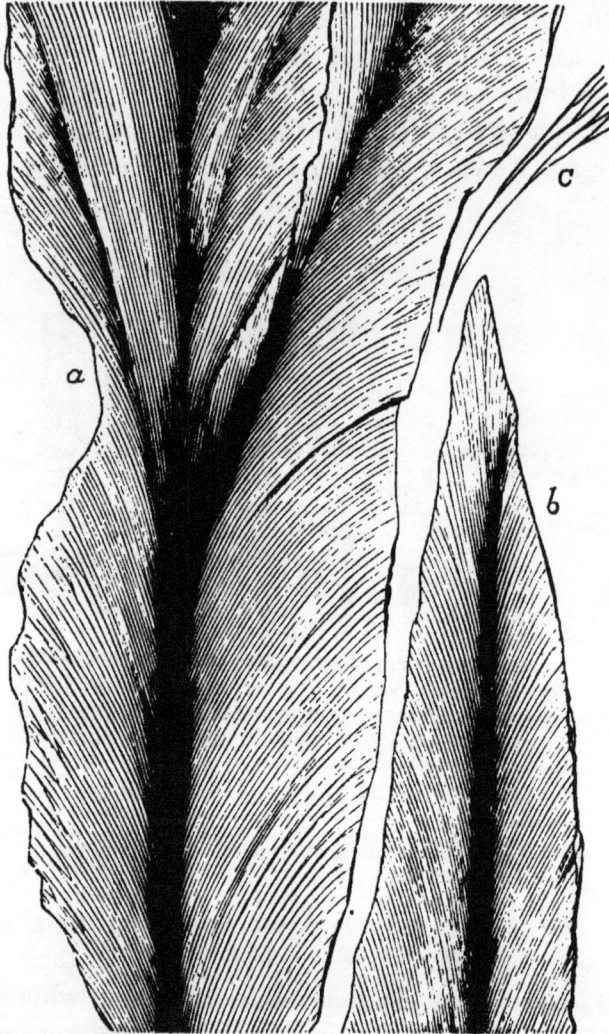

Fig. 26.—*Megalopteris Dawsoni*, Hartt (Erian, New Brunswick). *a*, Fragment of pinna. *b*, Point of pinnule. *c*, Venation. (The midrib is not accurately given in this figure.)

still continues in the modern tree-ferns of the tropics
and the southern hemisphere.   In one kind of tree-fern

stem from the Erian of New York, there is also a special arrangement for support, consisting of a series of peculiarly arranged radiating plates of scalariform vessels, not exactly like those of an exogenous stem, but doing duty for it (*Asteropteris*).*

Similar plants have been described from the Erian of Falkenberg, in Germany, and of Saalfeld, in Thuringia, by Goeppert and Unger, and are referred to ferns by the former, but treated as doubtful by the latter.† This peculiar type of treefern is apparently a precursor of the more exogenous type of *Heterangium*, recently described and referred to ferns by Williamson. Here, again, we have a mechanical contrivance now restricted to higher plants appropriated by these old cryptogams.

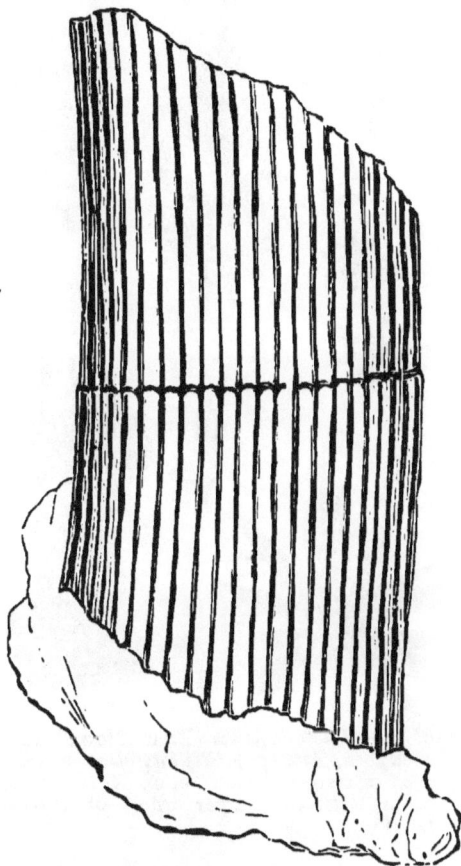

FIG. 27.—*Calamites radiatus* (Erian, New Brunswick).

The history of the ferns in geological time is remarkably different from that of the Lycopods; for while the

---

* "Journal of the Geological Society," London, 1881.

† "Sphenopteris Refracta," Goeppert; "Flora des Uebergangsgebirges." "Cladoxylon Mirabile," Unger; "Palæontologie des Thuringer Waldes."

latter have long ago descended from their pristine emi-
nence to a very humble place in nature, the former still,
in the southern hemisphere at least, retain their arboreal
dimensions and an-
cient dominance.

The family of
the *Equisetaceæ*, or
mare's-tails, was also
represented by large
species of *Calamites*
and by *Asterophyl-
lites* in the Erian ;
but, as its headquar-
ters are in the Car-
boniferous, we may
defer its considera-
tion till the next
chapter. (Figs. 27,
28.)

Passing over these
for the present, we
find that the flower-
ing plants are repre-
sented in the Erian
forests by at least
two types of Gym-
nosperms, that of
Taxineæ or yews,

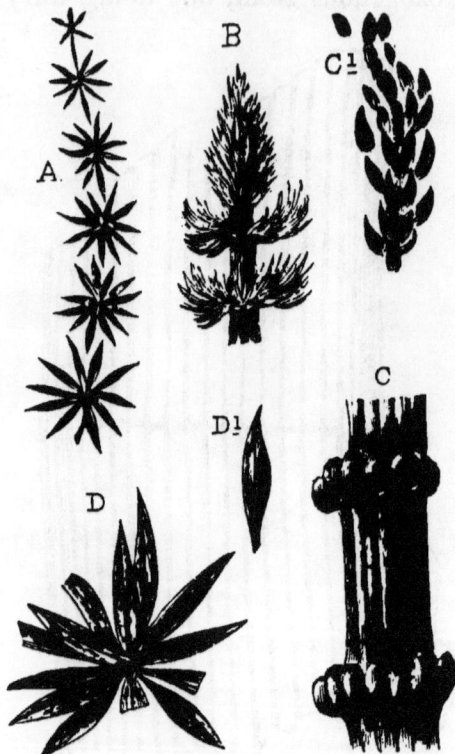

FIG. 28.—*Asterophyllites* (Erian, New Brunswick). A, *Asterophyllites latifolia.* B, Do.,
apex of stem (?) fruit. c, c¹, *A. scutigera.*
D, *A. latifolia*, larger whorl of leaves.
D¹, Leaf.

and an extinct family, that of the *Cordaites* (Figs. 30, 31).
The yew-trees are closely allied to the pines and spruces,
and are often included with them in the family of *Coniferæ.*
They differ, however, in the habit of producing berries or
drupe-like fruits instead of cones, and there is some
reason to believe that this was the habit of the Erian
trees of this group, though their wood in some instances resembles rather that of the Araucaria, or Nor-

folk Island pine, than that of the modern yews. These trees are chiefly known to us by their mineralised trunks, which are often found like drift-wood on modern sandbanks embedded in the Erian sandstones or limestones. It often shows its structure in the most perfect manner in specimens penetrated by calcite or silica, or by pyrite, and in which the original woody matter has

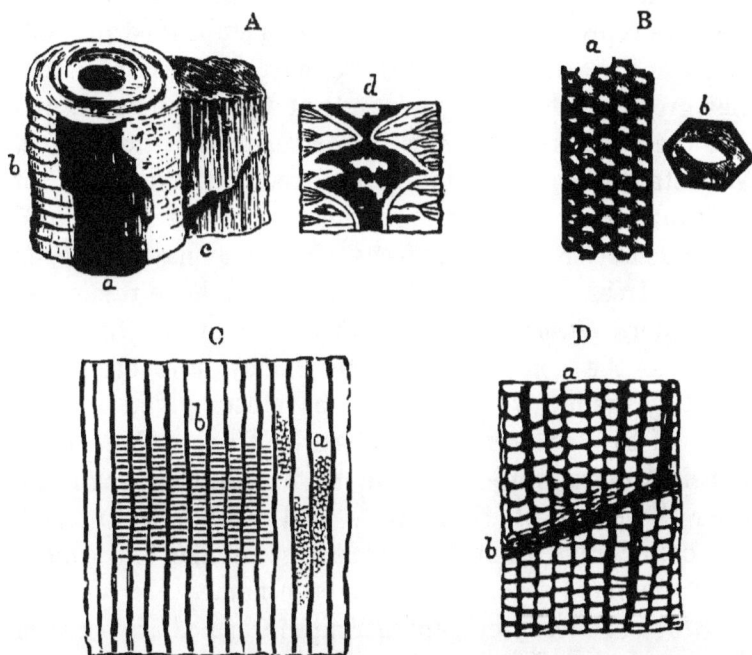

Fig. 29.—*Dadoxylon Ouangondianum*, an Erian conifer. A, Fragment showing Sternbergia pith and wood; *a*, medullary sheath; *b*, pith; *c*, wood; *d*, section of pith. B, Wood-cell; *a*, hexagonal areole; *b*, pore. C, Longitudinal section of wood, showing, *a*, areolation, and *b*, medullary rays. D, Transverse section, showing, *a*, wood-cells, and *b*, limit of layer of growth. (B, C, D, highly magnified.)

been resolved into anthracite or even into graphite. These trees have true woody tissues presenting that beautiful arrangement of pores or thin parts enclosed in cuplike discs, which is characteristic of the coniferous trees, and which is a great improvement on the barred tissue already referred to, affording a far more strong, tough, and durable wood, such as we have in our modern pines and yews (Fig. 29).

These primitive pines make their appearance in the Middle Erian, in various parts of America, as well as in Scotland and Germany, and they are represented by wood indicating the presence of several species. I have myself indicated and described five species from the Erian of Canada and the United States. From the fact that these trees are represented by drifted trunks embedded in sandstones and marine limestones, we may, perhaps, infer that they grew on the rising grounds of the Erian land, and that their trunks were carried by river-floods into the sea. No instance has yet certainly occurred of the discovery of their foliage or fruit, though there are some fan-shaped leaves usually regarded as ferns which may have belonged to such trees. These in that case would have resembled the modern *Gingko* of China, and some of the fruits referred to the genus *Cardiocarpum* may have been produced by them. Various names have been given to these trees. I have preferred that given by Unger, *Dadoxylon*, as being more non-committal as to affinities than the others.* Many of these trees had very long internal pith-cylinders, with curious transverse tubulæ, and which, when preserved separately, have been named *Sternbergia*.

Allied to these trees, and perhaps intermediate between them and the *Cycads*, were those known as *Cordaites* (Fig. 30), which had trunks resembling those of *Dadoxylon*, but with still larger Sternbergia piths and an internal axis of scalariform vessels, surrounded by a comparatively thin woody cylinder. Some of them have leaves over a foot in length, reminding one of the leaves of broad-leaved grasses or iridaceous plants. Yet their flowers and fruit seem to have been more nearly allied to the yews than to any other plants (Fig. 31). Their stems were less woody

---

* *Araucarites*, Goeppert; *Araucariozylon*, Kraus.

and their piths larger than in the true pines, and some of the larger-leaved species must have had thick, stiff branches. They are regarded as constituting a separate family, intermediate between pines and cycads, and, be-

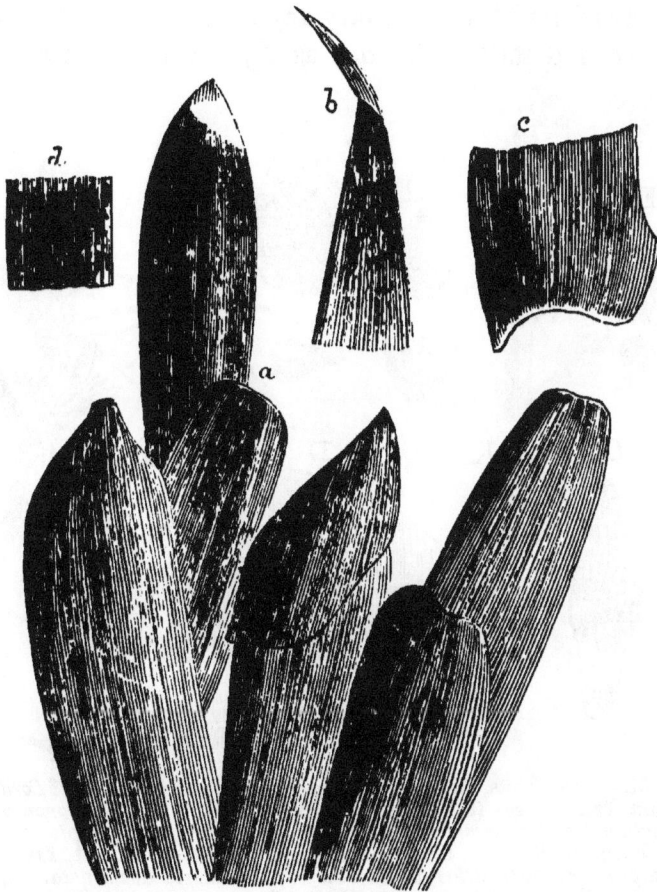

Fig. 30.—*Cordaites Robbii* (Erian, New Brunswick). *a*, Group of young leaves. *b*, Point of leaf. *c*, Base of leaf. *d*, Venation, magnified.

ginning in the Middle Devonian, they terminate in the Permian, where, however, some of the most gigantic species occur. In so far as the form and structure of the leaves, stems, and fruit are concerned, there is marvellously little difference between the species found in the

G

Erian and the Permian. They culminated, however, in
the Carboniferous period, and the coal-fields of southern
France have proved so far the richest in their remains.

Lastly, a single specimen, collected by Prof. James
Hall, of Albany, at Eighteen-mile Creek, Lake Erie, has
the structure of an ordinary angiospermous exogen, and
has been described by me as *Syringoxylon mirabile.**

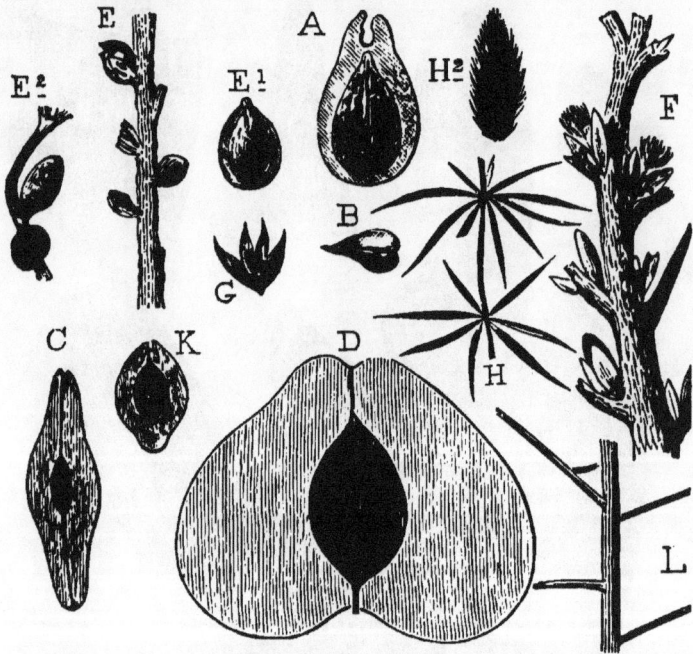

FIG. 31.—Erian fruits, &c., some gymnospermous, and probably of *Cordaites*
and Taxine trees (St. John, New Brunswick). A, *Cardiocarpum cornutum.* B, *Cardiocarpum acutum.* c, *Cardiocarpum Crampii.* D, *Cardiocarpum Baileyi.* E, *Trigonocarpum racemosum.* E¹, E², Fruits enlarged. F, *Antholithes Devonicus.* G, *Annularia acuminata.* H, *Asterophyllites acicularis.* H², Fruit of the same. K, *Cardiocarpum*
(? young of *A.*). L, *Pinnularia dispalans* (probably a root).

This unique example is sufficient to establish the fact of
the existence of such plants at this early date, unless some
accident may have carried a specimen from a later forma-

tion to be mixed with Erian fossils. It is to be observed, however, that the non-occurrence of any similar wood in all the formations between the Upper Erian and the Middle Cretaceous suggests very grave doubt as to the authenticity of the specimen. I record the fact, waiting further discoveries to confirm it. Of the character of the specimen which I have described I entertain no doubt.

We shall be better able to realise the significance and relations of this ancient flora when we have studied that of the succeeding Carboniferous. We may merely remark here on the fact that, in these forests of the Devonian and in the marshes on their margins, we find a wonderful expansion of the now modest groups of Rhizocarps and Lycopods, and that the flora as a whole belongs to the highest group of Cryptogams and the lowest of Phænogams, so that it has about it a remarkable aspect of mediocrity. Further, while there is evidence of some variety of station, there is also evidence of much equality of climate, and of a condition of things more resembling that of the insular climates of the temperate portions of the southern hemisphere than that of North America or Europe at present.

The only animal inhabitants of these Devonian woods, so far as known, were a few species of insects, discovered by Hartt in New Brunswick, and described by Dr. Scudder. Since, however, we now know that scorpions as well as insects existed in the Silurian, it is probable that these also occurred in the Erian, though their remains have not yet been discovered. All the known insects of the Erian woods are allies of the shad-flies and grasshoppers (*Neuroptera* and *Orthoptera*), or intermediate between the two. It is probable that the larvæ of most of them lived in water and fed upon the abundant vegetable matter there, or on the numerous minute crustaceans and worms. There were no land vertebrates, so far as known, but there were fishes (*Dipterus*, etc.), allied to the modern Barramunda or *Ceratodus* of Australia, and with teeth suited for grinding vegetable food. It is also possible that some of the smaller plate-covered fishes (Placoganoids, like *Pterichthys*) might have fed on vegetable matter, and, in any case, if they fed on lower animals, the latter must have subsisted on plants. I mention these facts to show that the superabundant vegetation of this age, whether aquatic or terrestrial, was not wholly useless to animals. It is quite likely, also, that we have yet much to learn of the animal life of the Erian swamps and woods.

NOTES TO CHAPTER III.

I.—CLASSIFICATION OF SPORANGITES.

It is, of course, very unsatisfactory to give names to mere fragments of plants, yet it seems very desirable to have some means of arranging them. With respect to the organisms described above, which were originally called by me *Sporangites*, under the supposition that they were Sporangia rather than spores, this name has so far been vindicated by the discovery of the spore-cases belonging to them, so that I think it may still be retained as a provisional name; but I would designate the whole as *Protosalviniæ*, meaning thereby plants with rhizocarpean affinities, though possibly when better understood belonging to different genera. We may under these names speak of their detached discs as macrospores and of their cellular envelopes as sporocarps. The following may be recognized as distinct forms:

1. *Protosalvinia Huronensis*, Dawson, *Syn., Sporangites Huronensis*, "Report on Erian Flora of Canada," 1871.—Macrospores, in the form of discs or globes, smooth and thick-walled, the walls penetrated by minute radiating pores. Diameter about one one-hundredth of an inch, or a little more, When *in situ* several macrospores are contained in a thin cellular sporocarp, probably globular in form. From the Upper Erian, and perhaps Lower Carboniferous shales of Kettle Point, Lake Huron, of various places in the State of Ohio, and in the shale boulders of the boulder clay of Chicago and vicinity. First collected at Kettle Point by Sir W. E. Logan, and

in Ohio by Prof. Edward Orton, and at Chicago by Dr. H. A. Johnson and Mr. B. W. Thomas, also in New York by Prof. J. M. Clarke.

The macrospores collected by Mr. Thomas from the Chicago clays and shales conform closely to those of Kettle Point, and probably belong to the same species. Some of them are thicker in the outer wall, and show the pores much more distinctly. These have been called by Mr. Thomas *S. Chicagoensis*, and may be regarded as a varietal form. Specimens isolated from the shale and mounted dry, show what seems to have been the hilum or scar of attachment better than those in balsam.

Sections of the Kettle Point shale show, in addition to the macrospores, wider and thinner shreds of vegetable matter, which I am inclined to suppose to be remains of the sporocarps.

2. *Protosalvinia (Sporangites) Braziliensis*, Dawson, "Canadian Record of Science," 1883.—Macrospores, round, smooth, a little longer than those of the last species, or about one seventy-fifth of an inch in diameter, enclosed in round, oval, or slightly reniform sporocarps, each containing from four to twenty-four macrospores. Longest diameter of sporocarps three to six millimetres. Structure of wall of sporocarps hexagonal cellular. Some sporocarps show no macrospores, and may possibly contain microspores. The specimens are from the Erian of Brazil. Discovered by Mr. Orville Derby. The formation, according to Mr. Derby, consists of black shales below, about three hundred feet thick, and containing the fucoid known as *Spirophyton*, and probably decomposed vegetable matter. Above this is chocolate and reddish shale, in which the well-preserved specimens of *Protosalvinia* occur. These beds are very widely distributed, and abound in *Protosalvinia* and *Spirophyton*.

3. *Protosalvinia (Sporangites) bilobata*, Dawson, "Canadian Record of Science," 1883.—Sporocarps, oval .or reniform, three to six millimetres in diameter, each showing two rounded prominences at the ends, with a depression in the middle, and sometimes a raised neck or isthmus at one side connecting the prominences. Structure of sporocarp cellular. Some of the specimens indicate that each prominence or tubercle contained several macrospores. At first sight it would be easy to mistake these bodies for valves of *Beyrichia*.

Found in the same formations with the last species, though, in so far as the specimens indicate, not precisely in the same beds. Collected by Mr. Derby.

4. *Protosalvinia Clarkei*, Dawson, *P. bilobata*, Clarke, "American Journal of Science."—Macrospores two-thirds to one millimetre in

diameter. One, two, or three contained in each sporocarp, which is cellular. The macrospores have very thick walls with radiating tortuous tubes. Unless this structure is a result of mineral crystallisation, these macrospores must have had very thick walls and must have resembled in structure the thickened cells of stone fruits and of the core of the pear, or the tests of the Silurian and Erian seeds known as *Pachytheca*, though on a smaller scale.

It is to be observed that bodies similar to these occur in the Boghead earthy bitumen, and have been described by Credner.

I have found similar bodies in the so-called "Stellar coal" of the coal district of Pictou, Nova Scotia, some layers of which are filled with them. They occur in groups or patches, which seem to be enclosed in a smooth and thin membrane or sporocarp. It is quite likely that these bodies are generically distinct from *Protosalvinia*.

5. *Protosalvinia punctata*, Newton, "Geological Magazine," New Series, December 2d, vol. ii.—Mr. Newton has named the discs found in the white coal and Tasmanite, *Tasmanites*, the species being *Tasmanites punctatus*, but as my name *Sporangites* had priority, I do not think it necessary to adopt this term, though there can be little doubt that these organisms are of similar character. The same remark may be made with reference to the bodies described by Huxley and Newton as occurring in the Better-bed coal.

In Witham's "Internal Structure of Fossil Vegetables," 1833, Plate XI, are figures of Lancashire cannel which shows *Sporangites* of the type of those in the Erian shales. Quekett, in his "Report on the Torbane Hill Mineral," 1854, has very well figured similar structures from the Methel coal and the Lesmahagow cannel coal. These are the earliest publications on the subject known to me; and Quekett, though not understanding the nature of the bodies he observed, holds that they are a usual ingredient in cannel coals.

II.—The Nature and Affinities of Ptilophyton.

(*Lycopodites Vanuxemii* of "Report on Devonian and Upper Silurian Plants," Part I., page 35. *L. plumula* of "Report on Lower Carboniferous Plants," page 24, Plate I., Figs. 7, 8, 9.) In the reports above referred to, these remarkable pinnate, frond-like objects were referred to the genus *Lycopodites*, as had been done by Goeppert in his description of the European species *Lycopodites pennæformis*, which is very near to the American Erian form. Since 1871, however, there have been many new specimens obtained, and very various opinions expressed as to their affinities. While Hall has named some of them *Plumalina*, and has regarded them as animal ctructures, allied to hydroids, Lesquereux has described some of the Carboniferous forms under the generic name *Trochophyllum*, which is, however, more appropriate to plants with verticillate leaves which are included in this genus. Before I had seen the publications of Hall and Lesquereux on the subject, I had in a paper on "Scottish Devonian Plants " * separated this group from the genus *Lycopodites*, and formed for it the genus *Ptilophyton*, in allusion to the feather-like aspect of the species. My reasons for this, and my present information as to the nature of these plants, may be stated as follows :

Schimper, in his "Palæontologie Vegetale " (possibly from inattention to the descriptions or want of access to specimens), doubts the lycopodiaceous character of species of *Lycopodites* described in my published papers on plants of the Devonian of America and in my Report of 1871. Of these, *L. Richardsoni* and *L. Matthewi* are undoubtedly very near to the modern genus *Lycopodium*. *L. Vanuxemii* is, I admit, more problematical ; but Schimper could scarcely have supposed it to be a fern or a fucoid allied to *Caulerpa* had he observed that both in my species and the allied *L. pennæformis* of Goeppert, which he does not appear to notice, the pinnules are articulated upon the stem, and leave scars where they have fallen off. When in Belfast in 1870, my attention was again directed to the affinities of these plants by finding in Prof. Thomson's collection a specimen from Caithness, which shows a plant apparently of this kind, with the same long narrow pinnæ or leaflets, attached, however, to thicker stems, and rolled up in a circinate manner. It seems to be a plant in vernation, and the parts are too much crowded and pressed together to admit of being accurately figured or described ; but I think I can scarcely be deceived as to its true nature. The circinate arrangement in this case would favour a relationship to ferns ; but some lycopodiaceous plants also roll themselves in this way, and so do the branches of the plants of the genus *Psilophyton*. (Fig. 17, *supra*.)

The specimen consists of a short, erect stem, on which are placed somewhat stout alternate branches, extending obliquely outward and then curving inward in a circinate manner. The lower ones appear to produce on their inner sides short lateral branchlets, and upon these, and also upon the curved extremities of the branches, are long, narrow, linear leaves placed in a crowded manner. The specimen is thus not a spike of fructification, but a young stem or branch in vernation, and which when unrolled would be of the form of those

---

* "Canadian Naturalist," 1878.

peculiar pinnate *Lycopodites* of which *L. Vanuxemii* of the American Devonian and *L. pennæformis* of the European Lower Carboniferous are the types, and it shows, what might have been anticipated from other specimens, that they were low, tufted plants, circinate in vernation. The short stem of this plant is simply furrowed, and bears no resemblance to a detached branch of *Lycopodites Milleri* which lies at right angles to it on the same slab. As to the affinities of the singular type of plants to which this specimen belongs, I may quote from my " Report on the Lower Carboniferous Plants of Canada," in which I have described an allied species, *L. plumula :*

" The botanical relations of these plants must remain subject to doubt, until either their internal structure or their fructification can be discovered. In the mean time I follow Goeppert in placing them in what we must regard as the provisional genus *Lycopodites*. On the one hand, they are not unlike the slender twigs of *Taxodium* and similar Conifers, and the highly carbonaceous character of the stems gives some colour to the supposition that they may have been woody plants. On the other hand, they might, so far as form is concerned, be placed with Algæ of the type of Brongniart's *Chondrites obtusus*, or the modern *Caulerpa plumaria*. Again, in a plant of this type from the Devonian of Caithness to which I have referred in a former memoir, the vernation seems to have been circinate, and Schimper has conjectured that these plants may be ferns, which seems also to have been the view of Shumard."

On the whole, these plants are allied to Lycopods rather than to ferns; and as they constitute a small but distinct group, known only, so far as I am aware, in the Lower Carboniferous and Erian or Devonian, they deserve a generic name, and I proposed for them in my " Paper on Scottish Devonian Plants," 1878, that of *Ptilophyton*, a name sufficiently distinct in sound from Psilophyton, and expressing very well their peculiar feather-like habit of growth. The genus was defined as follows :

" Branching plants, the branches bearing long, slender leaves in two or more ranks, giving them a feathered appearance; vernation circinate. Fruit unknown, but analogy would indicate that it was borne on the bases of the leaves or on modified branches with shorter leaves."

The Scottish specimen above referred to was named *Pt. Thomsoni*, and was characterised by its densely tufted form and thick branches. The other species known are: *Pt. pennæformis*, Goeppert, L. Carboniferous; *Pt. Vanuxemii*, Dawson, Devonian; *Pt. plumula*, Dawson, L. Carboniferous.

Shumard's *Filicites gracilis*, from the Devonian of Ohio, and Stur's *Pinites antecedens*, from the Lower Carboniferous of Silesia, may possibly belong to the same genus. The Scottish specimen referred to is apparently the first appearance of this form in the Devonian of Europe.

I have at a still later date had opportunities of studying considerable series of these plants collected by Prof. Williams, of Cornell University, and prepared a note in reference to them for the American Association, of which, however, only an abstract has been published. I have also been favoured by Prof. Lesquereux and Mr. Lacoe, of Pittston, with the opportunity of studying the specimens referred to *Trochophyllum*.

Prof. Williams's specimens occur in a dark shale associated with remains of land-plants of the genera *Psilophyton*, *Rhodea*, &c., and also marine shells, of which a small species of *Rhynchonella* is often attached to the stems of the *Ptilophyton*. Thus these organisms have evidently been deposited in marine beds, but in association with land-plants.

The study of the specimens collected by Prof. Williams develops the following facts: (1) The plants are not continuous fronds, but slender stems or petioles, with narrow, linear leaflets attached in a pinnate manner. (2) The pinnules are so articulated that they break off, leaving delicate transverse scars, and the lower parts of the stems are often thus denuded of pinnæ for the length of one or more inches. (3) The stems curve in such a manner as to indicate a circinate vernation. (4) In a few instances the fronds were observed to divide dichotomously toward the top; but this is rare. (5) There are no indications of cells in the pinnules; but, on the other hand, there is no appearance of fructification unless the minute granules which roughen some of the stems are of this nature. (6) The stems seem to have been lax and flexuous, and in some instances they seem to have grown on the petioles of ferns preserved with them in the same beds. (7) The frequency of the attachment of small brachiopods to the specimens of *Ptilophyton* would seem to indicate that the plant stood erect in the water. (8) Some of the specimens show so much carbonaceous matter as to indicate that the pinnules were of considerable consistency. All these characters are those rather of an aquatic plant than of an animal organism or of a land-plant.

The specimens communicated by Prof. Lesquereux and Mr. Lacoe are from the Lower Carboniferous, and evidently represent a different species with similar slender pitted stems, often partially denuded of pinnules below; but the pinnules are much broader and more distant. They are attached by very narrow bases, and apparently tend to lie on a plane, though they may possibly have been spirally arranged. On the same slabs are rounded sporangia or macrospores like those of *Lepidodendron*, but there is no evidence that these belonged to *Trochophyllum*. On the stems of this plant, however, there are small, rounded bodies apparently taking the places of some of the pinnules. These may possibly be spore-cases; but they may be merely imperfectly developed pinnules. Still the fact that similar small granules appear on the stems of the Devonian species, favours the idea that they may be organs of fructification.

The most interesting discovery, however, which results from the study of Mr. Lacoe's specimens, is that the pinnules were cylindrical and hollow, and probably served to float the plant. This would account for many of the peculiarities in the appearance and mode of occurrence of the Devonian *Ptilophyton*, which are readily explained if it is supposed to be an aquatic plant, attaching itself to the stems of submerged vegetable remains and standing erect in the water by virtue of its hollow leaves. It may well, however, have been a plant of higher organisation than the Algæ, though no doubt cryptogamous.

The species of *Ptilophyton* will thus constitute a peculiar group of aquatic plants, belonging to the Devonian and Lower Carboniferous periods, and perhaps allied to Lycopods and Pillworts in their organisation and fruit, but specially distinguished by their linear leaves serving as floats and arranged pinnately on slender stems. The only species yet found within the limits of Canada is *Pt. plumula*, found by Dr. Honeyman in the Lower Carboniferous of Nova Scotia; but as *Pt. Vanuxemii* abounds in the Erian of New York, it will no doubt be found in Canada also.

III.—TREE-FERNS OF THE ERIAN PERIOD.

As the fact of the occurrence of true tree-ferns in rocks so old as the Middle Erian or Devonian has been doubted in some quarters, the following summary is given from descriptions published in the "Journal of the Geological Society of London" (1871 and 1881), where figures of the species will be found:

Of the numerous ferns now known in the Middle and Upper Devonian of North America, a great number are small and delicate species, which were probably herbaceous; but there are other species which may have been tree-ferns. Little definite information, however, has, until recently, been obtained with regard to their habit of growth.

The only species known to me in the Devonian of Europe is the *Caulopteris Peachii* of Salter, figured in the " Quarterly Journal of the Geological Society " for 1858. The original specimen of this I had an opportunity of seeing in London, through the kindness of Mr. Etheridge, and have no doubt that it is the stem of a small arborescent fern, allied to the genus *Caulopteris*, of the coal formation.

In my paper on the Devonian of Eastern America (" Quarterly Journal of the Geological Society," 1862), I mentioned a plant found by Mr. Richardson at Perry, as possibly a species of *Megaphyton*, using that term to denote those stems of tree-ferns which have the leaf-scars in two vertical series; but the specimen was obscure, and I have not yet obtained any other.

More recently, in 1869, Prof. Hall placed in my hands an interesting collection from Gilboa, New York, and Madison County, New York, including two trunks surrounded by aërial roots, which I have described as *Psaronius textilis* and *P. Erianus*, in my " Revision of the Devonian Flora," read before the Royal Society.* In the same collection were two very large petioles, *Rhachiopteris gigantea* and *R. palmata*, which I have suggested may have belonged to tree-ferns.

My determination of the species of *Psaronius*, above mentioned, has recently been completely confirmed by the discovery on the part of Mr. Lockwood, of Gilboa, of the upper part of one of these stems, with its leaf-scars preserved and petioles attached, and also by some remarkable specimens obtained by Prof. Newberry, of New York, from the Corniferous limestone of Ohio, which indicate the existence there of three species of tree-ferns, one of them with aërial roots similar to those of the Gilboa specimens. The whole of these specimens Dr. Newberry has kindly allowed me to examine, and has permitted me to describe the Gilboa specimen, as connected with those which I formerly studied in Prof. Hall's collections. The specimens from Ohio he has himself named, but allows me to notice them here by way of comparison with the others. I shall add some notes on specimens found with the Gilboa ferns.

It may be further observed that the Gilboa specimens are from a bed containing erect stumps of tree-ferns, in the Chemung group of the Upper Devonian, while those from Ohio are from a marine limestone, belonging to the lower part of the Middle Devonian.

1. *Caulopteris Lockwoodi*, Dawson.—Trunk from two to three

---

* Abstract in " Proceedings of the Royal Society," May, 1870; also "Report on Erian Plants of Canada," 1871.

inches in diameter, rugose longitudinally. Leaf-scars broad, rounded above, and radiatingly rugose, with an irregular scar below, arranged spirally in about five ranks; vascular bundles not distinctly preserved. Petioles slender, much expanded at the base, dividing at first in a pinnate manner, and afterwards dichotomously. Ultimate pinnæ with remains of numerous, apparently narrow pinnules.

This stem is probably the upper part of one or other of the species of *Psaronius* found in the same bed (*P. Erianus*, Dawson, and *P. textilis*, Dawson).* It appears to have been an erect stem embedded *in situ* in sandstone, and preserved as a cast. The stem is small, being only two inches, or a little more, in diameter. It is coarsely wrinkled longitudinally, and covered with large leaf-scars, each an inch in diameter, of a horseshoe-shape. The petioles, five of which remain, separate from these scars with a distinct articulation, except at one point near the base, where probably a bundle or bundles of vessels passed into the petiole. They retain their form at the attachment to the stem, but a little distance from it they are flattened. They are inflated at the base, and somewhat rapidly diminish in size. The leaf-scars vary in form, and are not very distinct, but they appear to present a semicircular row of pits above, largest in the middle. From these there proceed downward a series of irregular furrows, converging to a second and more obscure semicircle of pits, within or below which is the irregular scar or break above referred to. The attitude and form of the petioles will be seen from Fig. 24, *supra*.

The petioles are broken off within a few inches of the stem; but other fragments found in the same beds appear to show their continuation, and some remains of their foliage. One specimen shows a series of processes at the sides, which seem to be the remains of small pinnæ, or possibly of spines on the margin of the petiole. Other fragments show the division of the frond, at first in a pinnate manner, and subsequently by bifurcation; and some fragments show remains of pinnules, possibly of fertile pinnules. These are very indistinct, but would seem to show that the plant approached, in the form of its fronds and the arrangement of its fructification, to the Cyclopterids of the subgenus *Aneimites*, one of which (*Aneimites Acadica*), from the Lower Carboniferous of Nova Scotia, I have elsewhere described as probably a tree-fern.† The

---

* Memoir on Devonian Flora, "Proceedings of the Royal Society," May, 1870.

† "Quarterly Journal of the Geological Society," 1860.

fronds were evidently different from those of *Archæopteris*,* a genus characteristic of the same beds, but of very different habit of growth. This accords with the fact that there is in Prof. Hall's collection a mass of fronds of *Cyclopteris (Archæopteris) Jacksoni*, so arranged as to make it probable that the plant was an herbaceous fern, producing tufts of fronds on short stems in the ordinary way. The obscurity of the leaf-scars may render it doubtful whether the plant above described should be placed in the genus *Caulopteris* or in *Stemmatopteris;* but it appears most nearly allied to the former. The genus is at present, of course, a provisional one; but I have thought it only justice to the diligent labours of Mr. Lockwood to name this curious and interesting fossil *Caulopteris Lockwoodi*.

I have elsewhere remarked on the fact that trunks, and petioles, and pinnules of ferns are curiously dissociated in the Devonian beds —an effect of water-sorting, characteristic of a period in which the conditions of deposition were so varied. Another example of this is, that in the sandstones of Gaspé Bay, which have not as yet afforded any example of fronds of ferns, there are compressed trunks, which Mr. Lockwood's specimens allow me at least to conjecture may have belonged to tree-ferns, although none of them are sufficiently perfect for description.

Mr. Lockwood's collection includes specimens of *Psaronius textilis;* and in addition to these there are remains of erect stems somewhat different in character, yet possibly belonging to the higher parts of the same species of tree-fern. One of these is a stem crushed in such a manner that it does not exhibit its form with any distinctness, but surrounded by smooth, cylindrical roots, radiating from it in bundles, proceeding at first horizontally, and then curving downward, and sometimes terminating in rounded ends. They resemble in form and size the aërial roots of *Psaronius Erianus;* and I believe them to be similar roots from a higher part of the stem, and some of them young and not prolonged sufficiently far to reach the ground. This specimen would thus represent the stem of *P. Erianus* at a higher level than those previously found. We can thus in imagination restore the trunk and crown of this once graceful tree-fern, though we have not the detail of its fronds. Mr. Lockwood's collections also contain a specimen of the large fern-petiole which I have named *Rhachiopteris punctata*. My original specimen was obtained by Prof. Hall from the same horizon in New York.

---

* The genus to which the well-known *Cyclopteris (Adiantites) Hibernicus* of the Devonian of Ireland belongs.

That of Mr. Lockwood is of larger size, but retains no remains of the frond. It must have belonged to a species quite distinct from *Caulopteris Lockwoodi*, but which may, like it, have been a tree-fern.

2. *Caulopteris antiqua*, Newberry.—This is a flattened stem, on a slab of limestone, containing Brachiopods, Trilobites, &c., of the Corniferous limestone. It is about eighteen inches in length, and three and a half inches in average breadth. The exposed side shows about twenty-two large leaf-scars arranged spirally. Each leaf, where broken off, has left a rough fracture; and above this is a semicircular impression of the petiole against the stem, which, as well as the surface of the bases of the petioles, is longitudinally striated or tuberculated. The structures are not preserved, but merely the outer epidermis, as a coaly film. The stem altogether much resembles *Caulopteris Peachii*, but is of larger size. It differs from *C. Lockwoodi* in the more elongated leaf-bases, and in the leaves being more remotely placed; but it is evidently of the same general character with that species.

3. *Caulopteris* (*Protopteris*) *peregrina*, Newberry.— This is a much more interesting species than the last, as belonging to a generic or subgeneric form not hitherto recognised below the Carboniferous, and having its minute structure in part preserved.

The specimens are, like the last, on slabs of marine limestone of the Corniferous formation, and flattened. One represents an upper portion of the stem with leaf-scars and remains of petioles; another a lower portion, with aërial roots. The upper part is three inches in diameter, and about a foot in length, and shows thirty leaf-scars, which are about three-fourths of an inch wide, and rather less in depth. The upper part presents a distinct rounded and sometimes double marginal line, sometimes with a slight depression in the middle. The lower part is irregular, and when most perfect shows seven slender vascular bundles, passing obliquely downward into the stem. The more perfect leaf-bases have the structure preserved, and show a delicate, thin-walled, oval parenchyma, while the vascular bundles show scalariform vessels with short bars in several rows, in the manner of many modern ferns. Some of the scars show traces of the hippocrepian mark characteristic of *Protopteris;* and the arrangement of the vascular bundles at the base of the scars is the same as in that genus, as are also the general form and arrangement of the scars. On careful examination, the species is indeed very near to the typical *P. Sternbergii*, as figured by Corda and Schimper.[*]

---

[*] Corda, "Beiträge," Pl. 48, copied by Schimper, Pl. 52.

The genus *Protopteris* of Sternberg, though the original species
(*P. punctata*) appears as a *Lepidodendron* in his earlier plate (Plate
4), and as a *Sigillaria* (*S. punctata*) in Brongniart's great work, is a
true tree-fern; and the structure of one species (*P. Cottai*) has been
beautifully figured by Corda. The species hitherto described are
from the Carboniferous and Permian.

The second specimen of this species represents a lower part of
the stem. It is thirteen inches long and about four inches in diameter, and is covered with a mass of flattened aërial roots lying parallel to each other, in the manner of the *Psaronites* of the coal-formation and of *P. Erianus* of the Upper Erian or Devonian.

4. *Asteropteris noveboracensis*, gen. and sp. n.—The genus *Asteropteris* is established for stems of ferns having the axial portion
composed of vertical radiating plates of scalariform tissue embedded
in parenchyma, and having the outer cylinder composed of elongated
cells traversed by leaf-bundles of the type of those of *Zygopteris*.

The only species known to me is represented by a stem 2·5 centimetres in diameter, slightly wrinkled and pitted externally, perhaps by traces of aërial roots which have perished. The transverse
section shows in the centre four vertical plates of scalariform or imperfectly reticulated tissue, placed at right angles to each other, and
united in the middle of the stem. At a short distance from the
centre, each of these plates divides into two or three, so as to form
an axis of from ten to twelve radiating plates, with remains of cellular tissue filling the angular interspaces. The greatest diameter of
this axis is about 1·5 centimetre. Exterior to the axis the stem consists of elongated cells, with somewhat thick walls, and more dense
toward the circumference. The walls of these cells present a curious
reticulated appearance, apparently caused by the cracking of the
ligneous lining in consequence of contraction in the process of car-
bonization. Embedded in this outer cylinder are about twelve vascular bundles, each with a dumb-bell-shaped group of scalariform
vessels enclosed in a sheath of thick-walled fibres. Each bundle is
opposite to one of the rays of the central axis. The specimen shows
about two inches of the length of the stem, and is somewhat bent,
apparently by pressure, at one end.

This stem is evidently that of a small tree-fern of a type, so
far as known to me, not before described,[*] and constituting a very
complex and symmetrical form of the group of Palæozoic ferns allied

---

[*] Prof. Williamson, to whom I have sent a tracing of the structure,
agrees with me that it is new.

to the genus *Zygopteris* of Schimper. The central axis alone has a curious resemblance to the peculiar stem described by Unger ("Devonian Flora of Thuringia") under the name of *Cladoxylon mirabile;* and it is just possible that this latter stem may be the axis of some allied plant. The large aërial roots of some modern tree-ferns of the genus *Angiopteris* have, however, an analogous radiating structure.

The specimen is from the collection of Berlin H. Wright, Esq., of Penn Yan, New York, and was found in the Portage group (Upper Erian) of Milo, New York, where it was associated with large petioles of ferns and trunks of *Lepidodendra*, probably *L. Chemungense* and *L. primœvum.*

The occurrence of this and other stems of tree-ferns in marine beds has recently been illustrated by the observation of Prof. A. Agassiz that considerable quantities of vegetable matter can be dredged from great depths in the sea on the leeward side of the Caribbean Islands. The occurrence of these trunks further connects itself with the great abundance of large petioles (*Rhachiopteris*) in the same beds, while the rarity of well-preserved fronds is explained by the coarseness of the beds, and also by the probably long maceration of the plant-remains in the sea-water.

In connection with this I may refer to the remarkable facts recently stated by Williamson [*] respecting the stems known as *Heterangium* and *Lyginodendron*. It would seem that these, while having strong exogenous peculiarities, are really stems of tree-ferns, thus placing this family in the same position of advancement with the Lycopods and *Equisetaceæ* of the Coal period.

IV.—ON ERIAN TREES OF THE GENUS DADOXYLON, UNGER.
(*Araucarites* OF GOEPPERT, *Araucarioxylon* OF KRAUS.)

Large woody trunks, carbonised or silicified, and showing wood-cells with hexagonal areoles having oval pores inscribed in them, occur abundantly in some beds of the Middle Erian of America, and constitute the most common kind of fossil wood all the way to the Trias. They have in the older formations, generally, several rows of pores on each fibre, and medullary rays composed of two or more series of cells, but become more simple in these respects in the Permian and Triassic series. The names *Araucarites* and *Araucarioxylon* are perhaps objectionable, inasmuch as they suppose affinities to *Araucaria* which may not exist. Unger's name, which is non-

---

[*] "Proceedings of the Royal Society," January 6, 1887.

committal, is therefore, I think, to be preferred. In my "Acadian Geology," and in my "Report on the Geology of Prince Edward Island," I have given reasons for believing that the foliage of some at least of these trees was that known as *Walchia*, and that they may have borne nutlets in the manner of Taxine trees (*Trigonocarpum*, &c.). Grand d'Eury has recently suggested that some of them may have belonged to *Cordaites*, or to plants included in that somewhat varied and probably artificial group.

The earliest discovery of trees of this kind in the Erian of America was that of Matthew and Hartt, who found large trunks. which I afterwards described as *Dadoxylon Ouangondianum*, in the Erian sandstone of St. John, New Brunswick, hence named by those geologists the "Dadoxylon sandstone." A little later, similar wood was found by Prof. Hall and Prof. Newberry in the Hamilton group of New York and Ohio, and the allied wood of the genus *Ormoxylon* was obtained by Prof. Hall in the Portage group of the former State. These woods proved to be specifically distinct from that of St. John, and were named by me *D. Halli, D. Newberryi,* and *Ormoxylon Erianum.* The three species of *Dadoxylon* agreed in having composite medullary rays, and would thus belong to the group *Palæoxylon* of Brongniart. In the case of *Ormoxylon* this character could not be very distinctly ascertained, but the medullary rays appeared to be simple.

I am indebted to Prof. J. M. Clarke, of Amherst College, Massachusetts, for some well-preserved specimens of another species from the Genesee shale of Canandaigua, New York. They show small stems or branches, with a cellular pith surrounded with wood of coniferous type, showing two to three rows of slit-formed, bordered pores in hexagonal borders. The medullary sheath consists of pseudo-scalariform and reticulated fibres; but the most remarkable feature of this wood is the structure of the medullary rays, which are very frequent, but short and simple, sometimes having as few as four cells superimposed. This is a character not before observed in coniferous trees of so great age, and allies this Middle Erian form with some Carboniferous woods which have been supposed to belong to *Cordaites* or *Sigillaria.* In any case this structure is new, and I have named the species *Dadoxylon Clarkii,* after its discoverer. The specimens occur, according to Prof. Clarke, in a calcareous layer which is filled with the minute shells of *Styliola fissurella* of Hall, believed to be a Pteropod; and containing also shells of *Goniatites* and *Gyroceras.* The stems found are only a few inches in diameter, but may be branches of larger trees.

It thus appears that we already know five species of Coniferous trees of the genus *Dadoxylon* in the Middle Erian of America, an interesting confirmation of the facts otherwise known as to the great richness and variety of this ancient flora. The late Prof. Goeppert informed me that he had recognised similar wood in the Devonian of Germany, and there can be no doubt that the fossil wood discovered by Hugh Miller in the Old Red Sandstone of Scotland, and described by Salter and McNab, is of similar character, and probably belongs to the genus *Dadoxylon*. Thus this type of Coniferous tree seems to have been as well established and differentiated into species in the Middle Devonian as in the succeeding Carboniferous.

I may here refer to the fact that the lower limit of the trees of this group coincides, in America, with the upper limit of those problematical trees which in the previous chapter I have named Protogens (*Nematophyton, Celluloxlyon,** *Nematoxylon* †), though *Aporoxylon* of Unger extends, in Thuringia, up to the Upper Devonian (Cypridina schists).

V.—Scottish Devonian Plants of Hugh Miller and others.
(Edinburgh Geological Society, 1877.)

Previously to the appearance of my descriptions of Devonian plants from North America, Hugh Miller had described forms from the Devonian of Scotland, similar to those for which I proposed the generic name *Psilophyton;* and I referred to these in this connection in my earliest description of that genus.‡ He had also recognised what seemed to be plants allied to Lycopods and Conifers. Mr. Peach and Mr. Duncan had made additional discoveries of this kind, and Sir J. Hooker and Mr. Salter had described some of these remains. More recently Messrs. Peach, Carruthers, and McNab have worked in this field, and still later * Messrs. Jack and Etheridge have summed up the facts and have added some that are new.

The first point to which I shall refer, and which will lead to the other matters to be discussed, is the relation of the characteristic *Lepidodendron* of the Devonian of eastern America, *L. Gaspianum,* to *L. nothum* of Unger and of Salter. At the time when I described this species I had not access to Scottish specimens of *Lepidodendron*

---

* "Journal of the Geological Society," May, 1881.

† *Ibid.*, vol. xix, 1863.

‡ "Journal of the Geological Society," London, 1859.

Ibid., 1877.

from the Devonian, but these had been well figured and described
by Salter, and had been identified with *L. nothum* of Unger, a species
evidently distinct from mine, as was also that figured and described
by Salter, whether identical or not with Unger's species. In 1870
I had for the first time an opportunity to study Scottish specimens
in the collection of Mr. Peach; and on the evidence thus afforded I
stated confidently that these specimens represented a species distinct
from *L. Gaspianum*, perhaps even generically so.* It differs from
L. Gaspianum in its habit of growth by developing small lateral
branches instead of bifurcating, and in its foliage by the absence or
obsolete character of the leaf-bases and the closely placed and somewhat appressed leaves. If an appearance of swelling at the end of a
lateral branch in one specimen indicates a strobile of fructification,
then its fruit was not dissimilar from that of the Canadian species
in its position and general form, though it may have differed in
details. On these grounds I declined to identify the Scottish species
with *L. Gaspianum*. The Lepidodendron from the Devonian of
Belgium described and figured by Crepin,† has a better claim to such
identification, and would seem to prove that this species existed in
Europe as well as in America. I also saw in Mr. Peach's collection
in 1870 some fragments which seemed to me distinct from Salter's
species, and possibly belonging to *L. Gaspianum*.‡

In the earliest description of *Psilophyton* I recognised its probable generic affinity with Miller's "dichotomous plants," with Salter's
"rootlets," and with Goeppert's *Haliserites Dechenianus*, and stated
that I had "little doubt that materials exist in the Old Red Sandstone of Scotland for the reconstruction of at least one species of
this genus." Since, however, Miller's plants had been referred to
coniferous roots, and to fucoids, and Goeppert's *Haliserites* was a
name applicable only to fucoids, and since the structure and fruit
of my plants placed them near to Lycopods, I was under the necessity of giving them a special generic name, nor could I with certainty affirm their specific identity with any European species. The
comparison of the Scottish specimens with woody rootlets, though
incorrect, is in one respect creditable to the acumen of Salter, as in
almost any state of preservation an experienced eye can readily per-
ceive that branchlets of *Psilophyton* must have been woody rather

---

* "Report on Devonian Plants of Canada," 1871.

† "Observations sur quelques Plantes Fossiles des dépôts Devoni-
ens."

‡ "Proceedings of the Geological Society of London," March, 1871.

than herbaceous, and their appearance is quite different from that of any true Algæ.

The type of *Psilophyton* is my *P. princeps*, of which the whole of the parts and structures are well known, the entire plant being furnished in abundance and *in situ* in the rich plant-beds of Gaspé. A second species, *P. robustius*, has also afforded well-characterised fructification. *P. elegans*, whose fruit appears as "oval scales," no doubt bore sac-like spore-cases resembling those of the other species, but in a different position, and perfectly flattened in the specimens procured. The only other Canadian species, *P. glabrum*, being somewhat different in appearance from the others, and not having afforded any fructification, must be regarded as uncertain.

The generic characters of the first three species may be stated as follows :

Stems dichotomous, with rudimentary subulate leaves, sometimes obsolete in terminal branchlets and fertile branches ; and in decorticated specimens represented only by punctiform scars. Young branches circinate. Rhizomata cylindrical, with circular root-areoles. Internal structure of stem, an axis of scalariform vessels enclosed in a sheath of imperfect woody tissue and covered with a cellular bark more dense externally. Fruit, naked sac-like spore-cases, in pairs or clusters, terminal or lateral.

The Scottish specimens conform to these characters in so far as they are known, but not having as yet afforded fruit or internal structure, they cannot be specifically determined with certainty. More complete specimens should be carefully searched for, and will no doubt be found.

In Belgium, M. Crepin has described a new species from the Upper Devonian of Condroz under the name *P. Condrusianum* (1875). It wants, however, some of the more important characters of the genus, and differs in having a pinnate ramification, giving it the aspect of a fern. In a later paper (1876) the author considers this species distinct from *Psilophyton*, and proposes for it a new generic name *Rhacophyton*.

The characters given by Mr. Carruthers, in his paper of 1873, for the species *P. Dechenianum*, are very few and general : "Lower branches short and frequently branching, giving the plant an oblong circumscription." Yet even these characters do not apply, so far as known, to Millor's fucoids or Salter's rootlets or Goeppert's *Halise-rites*. They merely express the peculiar mode of branching already referred to in Salter's *Lepidodendron nothum*. The identification of the former plants with the *Lepidodendron* and *Lycopodites*, indeed, rests only on mere juxtaposition of fragments, and on the slight resemblance of the decorticated ends of the branches of the latter plants to *Psilophyton*. It is contradicted by the obtuse ends of the branches of the *Lepidodendron* and *Lycopodites*, and by the apparently strobilaceous termination of some of them.

Salter's description of his *Lepidodendron nothum* is quite definite, and accords with specimens placed in my hands by Mr. Peach: "Stems half an inch broad, tapering little, branches short; set on at an acute angle, blunt at their terminations. Leaves in seven to ten rows, very short, not a line long, and rather spreading than closely imbricate." These characters, however, in so far as they go, are rather those of the genus *Lycopodites* than of *Lepidodendron*, from which this plant differs in wanting any distinct leaf-bases, and in its short, crowded leaves. It is to be observed that they apply also to Salter's *Lycopodites Milleri*, and that the difference of the foliage of that species may be a result merely of different state of preservation. For these reasons I am disposed to place these two supposed species together, and to retain for the species the name *Lycopodites Milleri*. It may be characterised by the description above given, with merely the modification that the leaves are sometimes nearly one-third of an inch long and secund (Fig. 17, *supra*, lower figure).

Decorticated branches of the above species may no doubt be mistaken for *Psilophyton*, but are nevertheless quite distinct from it, and the slender branching dichotomous stems, with terminations which, as Miller graphically states, are "like the tendrils of a pea," are too characteristic to be easily mistaken, even when neither fruit nor leaves appear. With reference to fructification, the form of *L. Milleri* renders it certain that it must have borne strobiles at the ends of its branchlets, or some substitute for these, and not naked spore-cases like those of *Psilophyton*.

The remarkable fragment communicated by Sir Philip Egerton to Mr. Carruthers,* belongs to a third group, and has, I think, been quite misunderstood. I am enabled to make this statement with some confidence, from the fact that the reverse or counterpart of Sir Philip's specimen was in the collection of Sir Wyville Thomson, and was placed by him in my hands in 1870. It was noticed in my paper on "New Devonian Plants," in the "Journal of the Geological Society of London," and referred to my genus *Ptilophyton*, as stated above under Section II., page 86 *et seq.*

---

* "Journal of Botany," 1873.

Mr. Salter described, in 1857,* fragments of fossil wood from the Scottish Devonian, having the structure of *Dadoxylon*, though very imperfectly preserved; and Prof. McNab has proposed † the generic name *Palæopitys* for another specimen of coniferous wood collected by Hugh Miller, and referred to by him in the "Testimony of the Rocks." From Prof. McNab's description, I should infer that this wood may, after all, be generically identical with the woods usually referred to *Dadoxylon* of Unger (*Araucarioxylon* of Kraus). The description, however, does not mention the number and disposition of the rows of pores, nor the structure of the medullary rays, and I have not been able to obtain access to the specimens themselves. I have described five species of *Dadoxylon* from the Middle and Upper Erian of America, all quite distinct from the Lower Carboniferous species. There is also one species of an allied genus, *Ormoxylon*. All these have been carefully figured, and it is much to be desired that the Scottish specimens should be re-examined and compared with them.

Messrs. Jack and Etheridge have given an excellent summary of our present knowledge of the Devonian flora of Scotland, in the Journal of the London Geological Society (1877). From this it would appear that species referable to the genera *Calamites*, *Lepidodendron*, *Lycopodites*, *Psilophyton*, *Arthrostigma*, *Archæopteris*, *Caulopteris*, *Palæopitys*, *Araucarioxylon*, and *Stigmaria* have been recognised.

The plants described by these gentlemen from the Old Red Sandstone of Callender, I should suppose, from their figures and descriptions, to belong to the genus *Arthrostigma*, rather than to *Psilophyton*. I do not attach any importance to the suggestions referred to by them, that the apparent leaves may be leaf-bases. Long leaf-bases, like those characteristic of *Lepidofloyos*, do not occur in these humbler plants of the Devonian. The stems with delicate "horizontal processes" to which they refer may belong to *Ptilophyton* or to *Pinnularia*.

In conclusion, I need scarcely say that I do not share in the doubts expressed by some British palæontologists as to the distinctness of the Devonian and Carboniferous floras. In eastern America, where these formations are mutually unconformable, there is, of course, less room for doubt than in Ireland and in western America, where they are stratigraphically continuous. Still, in passing

---

* "Journal of the London Geological Society."
† "Transactions of the Edinburgh Botanical Society," 1870.

from the one to the other, the species are for the most part different, and new generic forms are met with, and, as I have elsewhere shown, the physical conditions of the two periods were essentially different.*

It is, however, to be observed that since—as Stur and others have shown—*Calamites radiatus*, and other forms distinctively Devonian in America, occur in Europe in the Lower Carboniferous, it is not unlikely that the Devonian flora, like that of the Tertiary, appeared earlier in America. It is also probable, as I have shown in the "Reports" already referred to, that it appeared earlier in the Arctic than in the temperate zone. Hence an Arctic or American flora, really Devonian, may readily be mistaken for Lower Carboniferous by a botanist basing his calculations on the fossils of temperate Europe. Even in America itself, it would appear, from recent discoveries in Virginia and Ohio, that certain Devonian forms lingered longer in those regions than farther to the northeast; † and it would not be surprising if similar plants occurred in later beds in Devonshire or in the south of Europe than in Scotland. Still, these facts, properly understood, do not invalidate the evidence of fossil plants as to geological age, though errors arising from the neglect of them are still current.

VI.—GEOLOGICAL RELATIONS OF SOME PLANT-BEARING BEDS OF EASTERN CANADA. ("Report on Erian Plants," 1871.)

The Gaspé sandstones have been fully described by Sir W. E. Logan, in his "Report on the Geology of Canada," 1863. He there assigns to them a thickness of seven thousand and thirty-six feet, and shows that they rest conformably on the Upper Silurian limestones of the Lower Helderberg group (Ludlow), and are in their turn overlaid unconformably by the conglomerates which form the base of the Carboniferous rocks of New Brunswick. I shall add here merely a few remarks on points in their physical character connected with the occurrence of plants in them.

Prototaxites (*Nematophyton*) *Logani* and other characteristic Lower Erian plants occur in the base of the sandstones at Little Gaspé. This fact, along with the occurrence, as stated in my paper of 1863, of rhizomes of *Psilophyton* preserving their scalariform

---

* "Reports on Devonian Plants and Lower Carboniferous Plants of Canada."

† Andrews, "Palæontology of Ohio," vol. ii. ; Meek, "Fossil Plants from Western Virginia," Philosophical Society, Washington, 1875.

structure, in the upper part of the marine Upper Silurian lime-stones,* proves the flora of the Devonian rocks to have had its beginning at least in the previous geological period, and to charac-terise the lower as well as the upper beds of the Devonian series. In this connection I may state that, from their marine fossils, as well as their stratigraphical arrangement, Sir W. E. Logan and Mr. Billings regard the lower portions of the Gaspé sandstones as the equivalents of the Oriskany sandstone of New York. On the other hand, the great thickness of this formation, the absence of Lower Devonian fossils from its upper part, and the resemblance of the upper beds to those of the newer members of the Devonian else-where, render it probable that the Gaspé sandstones, though defi-cient in the calcareous members of the system, seen farther to the westward, represent the whole of the Devonian period.

The Gaspé sandstones, as their name imports, are predominantly arenaceous, and often coarsely so, the sandstones being frequently composed of large grains and studded with quartz-pebbles. Grey and buff are prevalent colours, but red beds also occur, more espe-cially in the upper portion. There are also interstratified shaly beds, sometimes occurring in groups of considerable thickness, and associated with fine-grained and laminated argillaceous sandstone, the whole having in many places the lithological aspect of the coal-measures. At one place, near the middle of the series, there is a bed of coal from one inch to three inches in thickness, associated with highly bituminous shales abounding in remains of plants, and also containing fragments of crustaceans and fishes (*Pterygotus, Ctenacanthus ?* &c.). The beds connected with this coal are grey sandstones and grey and dark shales, much resembling those of the ordinary coal formation. The coal is shining and laminated, and both its roof and floor consist of laminated bituminous shale with fragments of *Psilophyton*. It has no true under-clay, and has been, I believe, a peaty mass of rhizomes of *Psilophyton*. It occurs near Tar Point, on the south side of Gaspé Bay, a place so named from the occurrence of a thick dyke of trap holding petroleum in its cavities. The coal is of considerable horizontal extent, as in its line of strike a similar bed has been discovered on the Douglas River, about four miles distant. It has not been recognised on the north

---

* The marine fossils of these beds have been determined by Mr. Billings. They are Upper Silurian, with an intermixture of Lower Devo-nian in the upper part. Fragments of *Nematophyton* occur in beds of the same age in the Bay des Chaleurs, at Cape Bon Ami.

side of the bay, though we find there beds, probably on very nearly the same horizon, holding *Psilophyton in situ.*

As an illustration of one of the groups of shaly beds, and of the occurrence of roots of *Psilophyton*, I may give the following sectional list of beds seen near "Watering Brook," on the north shore of the bay. The order is descending:

			FT.	IN.
1.	Grey sandstones and reddish pebbly sandstone of great thickness			
2.	Bright-red shale................................		8	0
3.	Grey shales with stems of *Psilophyton*, very abundant but badly preserved.........................		0	5
4.	Grey incoherent clay, slickensided, and with many rhizomes and roots of *Psilophyton*.............		0	3
5.	Hard grey clay or shale, with fragments and roots of *Psilophyton*.................................		4	0
6.	Red shale.....................................		8	0
7.	Grey and reddish crumbling sandstone............			

Groups of beds similar to the above, but frequently much more rich in fossils, occur in many parts of the section, and evidently include fossil soils of the nature of under-clays, on which little else appears to have grown than a dense herbage of *Psilophyton*, along with plants of the genus *Arthrostigma*.

In addition to these shaly groups, there are numerous examples of beds of shale of small thickness included in coarse sandstones, and these beds often occur in detached fragments, as if the remnants of more continuous layers partially removed by currents of water. It is deserving of notice that nearly all these patches of shale are interlaced with roots or stems of *Psilophyton*, which sometimes project beyond their limits into the sandstone, as if the vegetable fibres had preserved the clay from removal. In short, these lines of patches of shale seem to be remnants of soils on which *Psilophyton* has flourished abundantly, and which have been partially swept away by the currents which deposited the sand. Some of the smaller patches may even be fragments of tough swamp soils interwoven with roots, drifted by the agency of the waves or possibly by ice; such masses are often moved in this way on the borders of modern swamps on the sea-coast.

The only remaining point connected with local geology to which I shall allude is the admirable facilities afforded by the Gaspé coast both for ascertaining the true geological relations of the beds, and for studying the Devonian plants, as distinctly exposed on large sur-

faces of rock.   On the coast of the river St. Lawrence, at Cape
Rozier and its vicinity, the Lower Silurian rocks of the Quebec
group are well exposed, and are overlaid unconformably by the massive Upper Silurian limestones of Cape Gaspé, which rise into cliffs
six hundred feet in height, and can be seen filled with their characteristic fossils on both sides of the cape.   Resting upon these, and
dipping at high angles toward Gaspé Bay, are the Devonian sandstones, which are exposed in rugged cliffs slightly oblique to their
line of strike, along a coast-line of ten miles in length, to the head
of the bay.   On the opposite side of the bay they reappear; and,
thrown into slight undulations by three anticlinal curves, occupy
a line, of coast fifteen miles in length.   The perfect manner in which
the plant-bearing beds are exposed in these fine natural sections may
serve to account for the completeness with which the forms and
habits of growth of the more abundant species can be described.

In the Bay des Chaleurs, similar rocks exist with some local
variations.   In the vicinity of Campbellton are calcareous and mag-
nesian breccia or agglomerate, hard shales, conglomerates and sandstones of Lower Devonian age.   The agglomerate and lower shales
contain abundant remains of fishes of the genera *Cephalaspis*, *Coc-
costeus*, *Ctenacanthus*, and *Homacanthus*, and also fragments of
Pterygotus.   The shales and sandstones abound in remains of *Psilophyton*, with which are *Nematophyton*, *Arthrostigma*, and *Leptophleum* of the same species found in the Lower Devonian of Gaspé
Bay.   These beds near Campbellton dip to the northward, and the
Restigouche River here occupies a synclinal, for on the opposite side,
at Bordeaux Quarry, there are thick beds of grey sandstone dipping
to the southward, and containing large silicified trunks of *Prototaxites*, in addition to *Psilophyton*.   These beds are all undoubtedly
Lower Erian, but farther to the eastward, on the north side of the
river, there are newer and overlying strata.   These are best seen at
Scaumenac Bay, opposite Dalhousie, between Cape Florissant and
Maguacha Point, where they consist of laminated and fine-grained
sandstone, with shales of grey colours, but holding some reddish beds
at top, and overlaid unconformably by a great thickness of Lower
Carboniferous red conglomerate and sandstone.   In these beds numerous fossil fishes have been found, among which Mr. Whiteaves
recognises species of *Pterichthys*, *Glyptolepis*, *Cheirolepis*, &c.   With
these are found somewhat plentifully four species of fossil ferns, all
of Upper Erian types, of which one is peculiar to this locality; but
the others are found in the Upper Erian of Perry, in Maine, or in
the Catskill group of New York.

In order that distinct notions may be conveyed as to the geological horizons of the species, I may state that the typical Devonian or Erian series of Canada and New York may be divided in descending order into—1. The Chemung group, including the Chemung and Portage sandstones and shales. 2. The Hamilton group, including the Genesee, Hamilton, and Marcellus shales. 3. The Corniferous limestone and its associated beds. 4. The Oriskany sandstone. As the Corniferous limestone, which is the equivalent of the Lower Carboniferous limestone in the Carboniferous period, is marine, and affords scarcely any plants, we may, as is usually done for like purposes in the Carboniferous, group it with the Oriskany under the name Lower Erian. The Hamilton rocks will then be Middle Erian, and the Chemung group Upper Erian. In the present state of our knowledge, the series may be co-ordinated with the rocks of Gaspé, New Brunswick, and Maine, as in the following table:

Subdivisions.	New York and Western Canada.	Gaspé and Bay des Chaleurs.	Southern New Brunswick.	Coast of Maine.
Upper Devonian or Erian.	Chemung Group.	Upper Sandstones. Long Cove, &c. Scauminac Beds.	Mispec Group. Shale, Sandstone, and Conglomerate.	Perry Sandstones.
Middle Devonian or Erian.	Hamilton Group.	Middle Sandstones. Bois Brulé, Cape Oiseau, &c.	Little R. Group (including Cordaite Shales and Dadoxylon Sandstone).	
Lower Devonian or Erian.	Corniferous and Oriskany groups.	Lower Sandstones. Gaspé Basin, Little Gaspé, &c. Campbellton Beds.	Lower Conglomerates, &c.	

It may be proper, before closing this note, to state the reasons which have induced me to suggest in the following pages the use of the term "ERIAN," as equivalent to "Devonian," for the great system of formations intervening between the Upper Silurian and the Lower Carboniferous in America. I have been induced to adopt this course by the following considerations: 1. The great area of undisturbed and unaltered rocks of this age, including a thickness in some places of eighteen thousand feet, and extending from east to west through the Northern States of the Union and western Canada for nearly seven hundred miles, while it spreads from north to south from the northern part of Michigan far into the Middle States, is undoubtedly the most important Devonian area now known to geologists. 2. This area has been taken by all American geologists as their typical Devonian region. It is rich in fossils, and these have been thoroughly studied and admirably illustrated by the New York and Canadian Surveys. 3. The rocks of this area surround the basin of Lake Erie, and were named, in the original reports of the New York Survey, the "*Erie Division*." 4. Great difficulties have been experienced in the classification of the European Devonian, and the uncertainties thus arising have tended to throw doubt on the results obtained in America in circumstances in which such difficulties do not occur.

These reasons are, I think, sufficient to warrant me in holding the great *Erie Division* of the New York geologists as the typical representative of the rocks deposited between the close of the Upper Silurian and the beginning of the Carboniferous period, and to use the term Erian as the designation of this great series of deposits as developed in America, in so far at least as their flora is concerned. In doing so, I do not wish to introduce a new name merely for the sake of novelty; but I hope to keep before the minds of geologists the caution that they should not measure the Erian formations of America, or the fossils which they contain, by the comparatively depauperated representatives of this portion of the geological scale in the Devonian of western Europe.

VII.—On the Relations of the so-called "Ursa Stage" of Bear Island with the Palæozoic Flora of North America.

The following note is a verbatim copy of that published by me in 1873, and the accuracy of which has now been vindicated by the recent observations of Nathorst:

The plants catalogued by Dr. Heer, and characterising what he calls the "Ursa Stage," are in part representatives of those of the American flora which I have described as the "Lower Carboniferous Coal-Measures" (Subcarboniferous of Dana), and whose characteristic species, as developed in Nova Scotia, I noticed in the "Journal of the Geological Society" in 1858 (vol. xv.). Dr. Heer's list, however, includes some Upper Devonian forms; and I would suggest that either the plants of two distinct beds, one Lower Carboniferous and the other Upper Devonian, have been near to or in contact with each other and have been intermixed, or else that in this high northern latitude, in which (for reasons stated in my "Report on the Devonian Flora" *) I believe the Devonian plants to have originated, there was an actual intermixture of the two floras. In America, at the base of the Carboniferous of Ohio, a transition of this kind seems to occur; but elsewhere in northeastern America the Lower Carboniferous plants are usually unmixed with the Devonian.

Dr. Heer, however, proceeds to identify these plants with those of the American Chemung, and even with those of the Middle Devonian of New Brunswick, as described by me—a conclusion from which I must altogether dissent, inasmuch as the latter belong to beds which were disturbed and partially metamorphosed before the deposition of the lowest Carboniferous or "Subcarboniferous" beds.

Dr. Heer's error seems to have arisen from want of acquaintance with the rich flora of the Middle Devonian, which, while differing in species, has much resemblance in its general facies, and especially in its richness in ferns, to that of the coal-formation.

To geologists acquainted with the stratigraphy and the accompanying animal fossils, Dr. Heer's conclusions will of course appear untenable; but they may regard them as invalidating the evidence of fossil plants; and for this reason it is, I think, desirable to give publicity to the above statements.

I consider the British equivalent of the lower coal-measures of eastern America to be the lower limestone shales, the *Tweedian group* of Mr. Tate (1858), but which have sometimes been called the "Calciferous Sandstone" (a name preoccupied for a Cambrian group in America). This group does not constitute "beds of passage" to the Devonian, more especially in eastern America, where the lower coal-formation rests unconformably on the Devonian, and is broadly distinguished by its fossils.

The above notes would not have been extended to so great length, but for the importance of the Erian flora as the precursor of that of the Carboniferous, and the small amount of attention hitherto given to it by geologists and botanists.

---

* "Geological Survey of Canada," 1871.

CHAPTER IV.

THE CARBONIFEROUS FLORA—CULMINATION OF THE ACROGENS—FORMATION OF COAL.

ASCENDING from the Erian to the Carboniferous system, so called because it contains the greatest deposits of anthracite and bituminous coal, we are still within the limits of the Palæozoic period. We are still within the reign of the gigantic club-mosses, cordaites, and taxine pines. At the close of the Erian there had been over the whole northern hemisphere great changes of level, accompanied by active volcanic phenomena, and under these influences the land flora seems to have much diminished. At length all the old Erian species had become extinct, and their place was supplied by a meagre group of lycopods, ferns, and pines of different species from those of the preceding Erian. This is the flora of the Lower Carboniferous series, the Tweedian of England, the Horton series of Nova Scotia, the lower coal-measures of Virginia, the culm of Germany. But the land again subsided, and the period of the marine limestone of the Lower Carboniferous was introduced. In this the older flora disappeared, and when the land emerged we find it covered with the rich flora of the coal-formation proper, in which the great tribes of the lycopods and cordaites attained their maxima, and the ferns were continued as before, though under new generic and specific forms.

There is something very striking in this succession of a new plant world without any material advance. It is like passing in the modern world from one district to another, in which we see the same forms of life, only represented by distinct though allied species. Thus, when the voyager crosses the Atlantic from Europe to America, he meets with pines, oaks, birches, poplars, and beeches of the same genera with those he had left behind ; but the species are distinct. It is something like this that meets us in our ascent into the Carboniferous world of plants. Yet we know that this is a succession in time, that all our old Erian friends are dead and buried long ago, and that these are new forms lately introduced (Fig. 32).

Conveying ourselves, then, in imagination forward to the time when our greatest accumulations of coal were formed,

a b c d e f g

Fig. 32. — Foliage from the coal-formation. *a, Alethopteris lonchitica,* fern (Moose River). *b, Sphenophyllum Schlotheimii* (Pictou). *c, Lepidodendron binerve* (Sydney). *d, Asterophyllites foliosa* (?) (Sydney). *e, Cordaites* (Joggins). *f, Neuropteris rarinervis,* fern (Sydney). *g, Odontopteris subcuneata,* fern (Sydney).

and fancying that we are introduced to the American or European continent of that period, we find ourselves in a new and strange world. In the Devonian age, and even in the succeeding Lower Carboniferous, there was in the interior of America a wide inland sea, with forest belts clinging to its sides or clothing its islands. But in the coal period this inland sea had given

place to vast swampy flats, and which, instead of the oil-
bearing shales of the Erian, were destined to produce
those immense and wide-spread accumulations of vegetable matter which constitute our present beds of bituminous and anthracite coal. The
atmosphere of these great swamps
is moist and warm. Their vegetation is most exuberant, but of
forms unfamiliar to modern eyes,
and they swarm with insects,
millepedes, and scorpions, and
with batrachian reptiles large
and small, among which we look
in vain for representatives of the
birds and beasts of the present
day.

Prominent among the more
gigantic trees of these swampy
forests are those known to us as
Sigillariæ (Fig. 33). They have
tall, pillar-like trunks, often several feet in diameter, ribbed like
fluted columns, but in the reverse way, and spreading at the
top into a few thick branches,
which are clothed with long,
grass-like leaves. They resemble in some respects the Lepidodendra of the Erian age, but
are more massive, with ribbed instead of scaly trunks, and longer
leaves. If we approach one of

Fig. 33.—*Sigillariæ*, restored.
A, *Sigillaria Brownii*.
B, *Sigillaria elegans*.

them more closely, we are struck with the regular ribs of
its trunk, dotted with rows of scars of fallen leaves, from
which it receives its name *Sigillaria*, or seal-tree (Figs.
34–37). If we cut into its stem, we find that, instead of

the thin bark and firm wood with which we are familiar
in our modern trees, it has a hard external rind, then a
great thickness of cellular matter with rope-like bands of
fibres, constituting an inner bark, while in the centre is
a firm, woody axis of comparatively small diameter, and

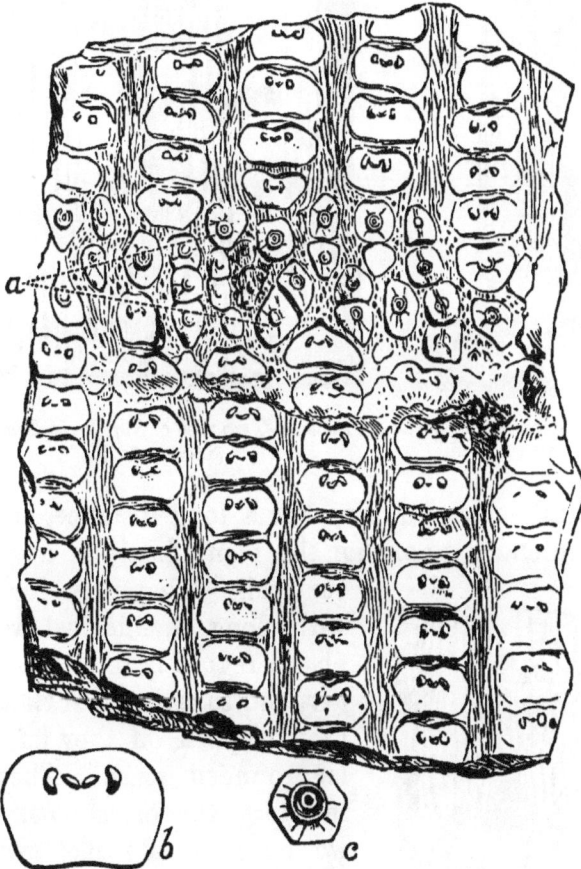

Fig. 34.—*Sigillaria Lorwayana*, Dawson. *a*, Zones of fruit-scars. *b*, Leaf-
scar enlarged. *c*, Fruit-scar enlarged. See appended note.

somewhat intermediate in its structures between that of
the Lepidodendra and those of the cycads and the taxine
conifers. Thus a great stem, five feet in diameter, may
consist principally of cellular and bast fibres with very
little true woody matter. The roots of this tree are

perhaps its most singular feature. They usually start from the stem in four main branches, then regularly bifurcate several times, and then run out into great

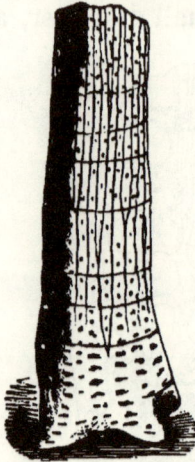

FIG. 35.—Stem of *Sigillaria Brownii*, reduced.

FIG. 36.—Two ribs of *Sigillaria Brownii*. Natural size.

cylindrical cables, running for a long distance, and evidently intended to anchor the plant firmly in a soft and oozy soil. They were furnished with long, cylindrical rootlets placed regularly in a spiral manner, and so articulated that when they dropped off they left regular rounded scars. They are, in short, the *Stigmariæ*, which we have already met with in the Erian (Figs. 38, 39). In Fig. 33 I have endeavoured to restore these strange trees. It is not wonderful that such plants have caused much botanical controversy. It was long before botanists could be convinced that

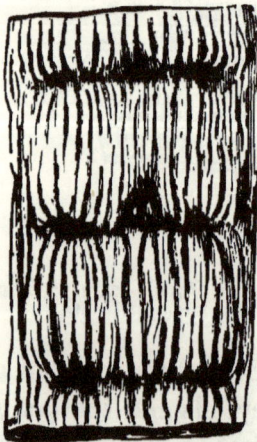

FIG. 37.—Portion of lower part of stem of *S. Brownii*. Natural size.

their roots are properly roots at all, and not stems
of some aquatic plant. Then the structure of their

Fig. 38.—*Stigmaria* root, seen from above, showing its regular divisions.
From "Acadian Geology."

stems is most puzzling, and their fruit is an enigma,
for while some have found connected with them cones
supposed to resemble those of
lycopods, others attribute to
them fruits like those of yew-
trees. For years I have been
myself gathering materials from
the rich coal-formation deposits
of Nova Scotia in aid of the
solution of these questions, and
in the mean time Dr. Williamson, of Manchester, and Renault
and other botanists in France,
have been amassing and studying stores of specimens, and it
is still uncertain who may finally be the fortunate discoverer
to set all controversies at rest.

Fig. 39.—Portion of bark of
Stigmaria, showing scars
of attachment of rootlets.

My present belief is,
that the true solution consists in the fact that there are
many kinds of *Sigillariæ*. While in the modern forests

of America and Europe the species of any of our ordinary
trees, as oaks, birches, or maples, may almost be counted
on one's fingers, Schimper in his vegetable palæontology
enumerates about eighty species of Carboniferous *Sigillariæ ;* and while on the one hand many of these are so
imperfectly known that they may be regarded as uncertain, on the other hand many species must yet remain to
be discovered.* Now, in so vast a number of species
there must be a great range of organisation, and, indeed,
it has already been attempted to subdivide them into
several generic groups. The present state of the question
appears to me to be this, that in these *Sigillariæ* we have
a group divisible into several forms, some of which will
eventually be classed with the Lepidodendra as lycopods,
while others will be found to be naked-seeded phænogams, allied to the pines and cycads, and to a remarkable
group of trees known as *Cordaites,* which we must shortly
notice.

Before considering other forms of Carboniferous vegetation, let us glance at the accumulation of coal, and the
agency of the forests of *Sigillariæ* therein. Let us imagine, in the first instance, such trees as those represented
in the figures, growing thickly together over vast swampy
flats, with quantities of undergrowth of ferns and other
plants beneath their shade, and accumulating from age to
age in a moist soil and climate a vast thickness of vegetable mould and trunks of trees, and spores and spore-
cases, and we have the conditions necessary for the growth
of coal. Many years ago it was observed by Sir William
Logan that in the coal-field of South Wales it was the
rule with rare exceptions that, under every bed of coal,
there is a bed of clay filled with roots of the *Stigmaria,*
already referred to as the root of *Sigillaria.* This dis-

---

* In a recent memoir (Berlin, 1887) Stur has raised the number of
species in one subdivision of the *Sigillariæ* (the *Favulariæ*) to forty-
seven!

covery has since been extended to all the coal-fields of
Europe and America, and it is a perfectly conclusive fact
as regards the origin of coal. Each of these "underclays," as they are called, must, in fact, have been a soil
on which grew, in the first instance, Sigillariæ and other
trees having stigmaria-roots. Thus, the growth of a
forest of *Sigillariæ* was the first step toward the accumulation of a bed of coal. More than this, in some of the
coarser and more impure coals, where there has been
sufficient earthy matter to separate and preserve impressions of vegetable forms, we can see that the mass of the
coal is made up of flattened *Sigillariæ*, mixed with vege-

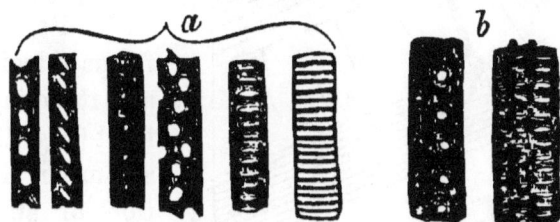

Fig. 40.—Vegetable tissues from coal. *a, Sigillaria* and *Cordaites.*
b, Calamodendron.

table *débris* of all kinds, including sometimes vast quantities of lepidodendroid spores, and the microscopic study
of the coal gives similar results (Fig. 40). Further, on
the surfaces of many coals, and penetrating the shales or
sandstones which form their roofs, we find erect stumps
of sigillaria and other trees, showing that the accumulation of the coal terminated as it had begun, by a forest-
growth. I introduce here a section of a few of the numerous beds of coal exposed in the cliffs of the South
Joggins, in Nova Scotia, in illustration of these facts.
We can thus see how in the slowly subsiding areas of the
coal-swamps successive beds of coal were accumulated,
alternating with beds of sandstone and shale (Figs. 41,
42). For other details of this kind I must refer to
papers mentioned in the sequel.

Returning to the more special subject of this work, I may remark that the lepidodendroid trees and the ferns, both the arborescent and herbaceous kinds, are even more richly represented in the Carboniferous than in the preceding Erian.   I must, however, content myself with merely introducing a few representatives of some of the more common kinds, in an appended note, and here give a figure of a well-known Lower Carboniferous lepidodendron, with its various forms of leaf-bases, and its foliage and fruit (Fig. 43), and a similar illustration of an allied generic form, that known as *Lepidophloios* * (Fig. 44).

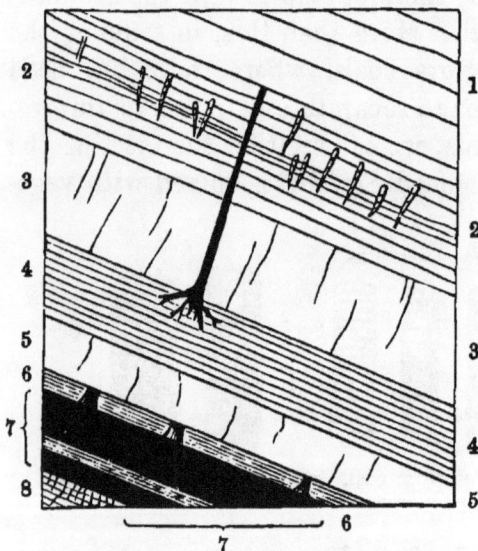

FIG. 41.—Beds associated with the main coal (S. Joggins, Nova Scotia). 1, Shale and sandstone—plants with *Spirorbis* attached; rainmarks (?). (2, Sandstone and shale, eight feet—erect *Calamites*; 3, Gray sandstone, seven feet; 4, Gray shale, four feet—an erect coniferous (?) tree, rooted on the shale, passes up through fifteen feet of the sandstones and shale.) 5, Gray sandstone, four feet. 6, Gray shale, six inches—prostrate and erect trees, with rootlets, leaves, *Naiadites*, and *Spirorbis* on the plants. 7, Main coal-seam, five feet of coal in two seams. 8, Underclay, with rootlets.

Another group which claims our attention is that of the *Calamites*. These are tall, cylindrical, branchless stems, with whorls of branchlets, bearing needle-like leaves and spreading in stools from the base, so as to form dense thickets, like Southern cane-brakes (Fig. 46). They bear, in habit of growth and fructification, a close

---

* For full descriptions of these, see " Acadian Geology."

relation to our modern equisetums, or mare's-tails, but, as in other cases we have met with, are of gigantic size and comparatively complex structure. Their stems, in cross-section, show radiating bundles of fibres, like those of exogenous woods, yet the whole plan of structure presents some curious resemblances to the stems of their humble successors, the modern mare's-tails. It would seem, from the manner in which dense brakes of these *Calamites* have been preserved in the coal-formation of Nova Scotia, that they spread over low and occasionally inundated flats, and formed fringes on the seaward sides of the great Sigillaria forests. In this way they no doubt contrib-

Fig. 42.—Erect *Sigillaria*, standing on a coal-seam (S. Joggins, Nova Scotia).

uted to prevent the invasion of the areas of coal accumulation by the muddy waters of inundations, and thus, though they may not have furnished much of the material of coal, they no doubt contributed to its purity. Many beautiful plants of the genera *Asterophyllites* and *Annularia* are supposed to have been allied to the *Calamites*, or to have connected them with the *Rhizocarps*. The stems and fruit of these plants have strong points of resemblance to those of *Sphenophyllum*, and the leaves are broad, and not narrow and angular like those of the true *Calamites* (Fig. 45).

No one has done more than my friend Dr. William-

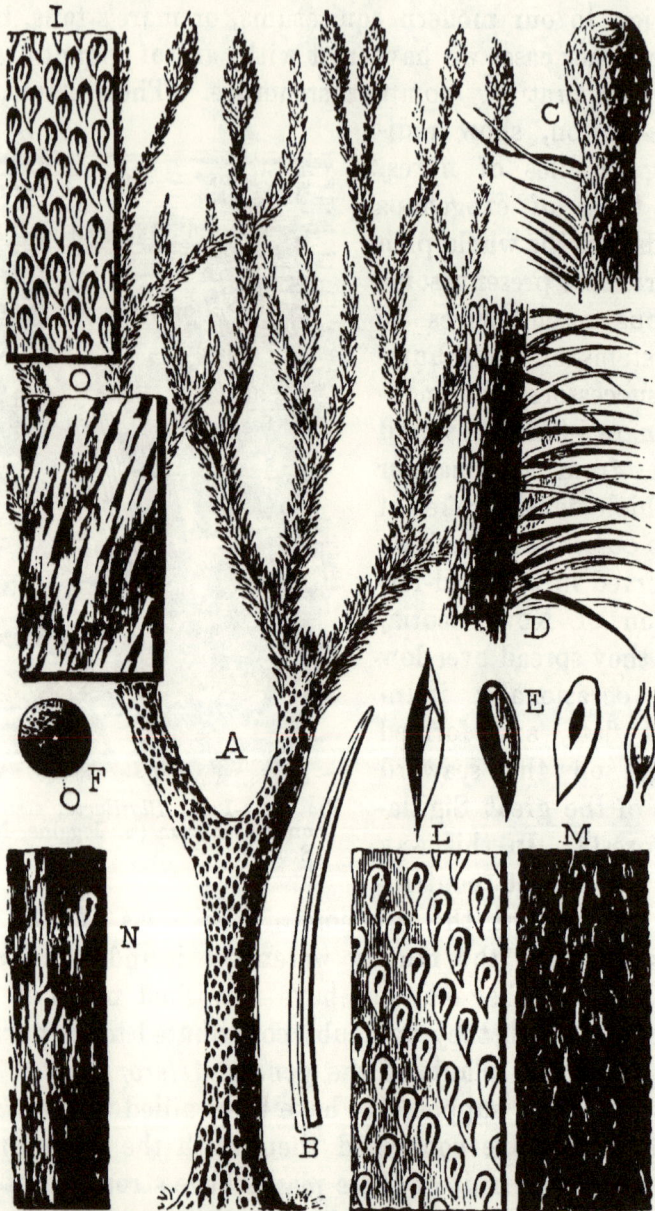

FIG. 43.—*Lepidodendron corrugatum*, Dawson, a tree characteristic of the Lower Carboniferous. A, Restoration. B, Leaf, natural size. C, Cone and branch. D, Branch and leaves. E, Various forms of leaf-areoles. F, *Sporangium*. I, L, M, Bark, with leaf-scars. N, Bark, with leaf-scars of old stem. O, Decorticated stem (*Knorria*).

FIG. 44.—*Lepidophloios Acadianus*, Dawson, a lepidodendroid tree of the coal-formation. A, Restoration. B, Portion of bark (two thirds natural size). C, Ligneous surface of the same. F, Cone (two thirds natural size). G, Leaf (natural size). K, Portion of woody cylinder, showing outer and inner series of vessels magnified. L, Scalariform vessels (highly magnified). M, Various forms of leaf-scars and leaf-bases (natural size).

son, of Manchester, to illustrate the structure of Calamites, and he has shown that these plants, like other
cryptogams of the Carboniferous, had mostly stems with
regular fibrous wedges, like those of exogens.   The
structure of the stem is, indeed, so complex, and differs
so much in different stages of growth, and different states
of preservation, that we are in danger of falling into the
greatest confusion in classifying these plants. Sometimes
what we call a Calamite is a mere cast of its pith showing
longitudinal striæ and constrictions at the nodes.   Some-

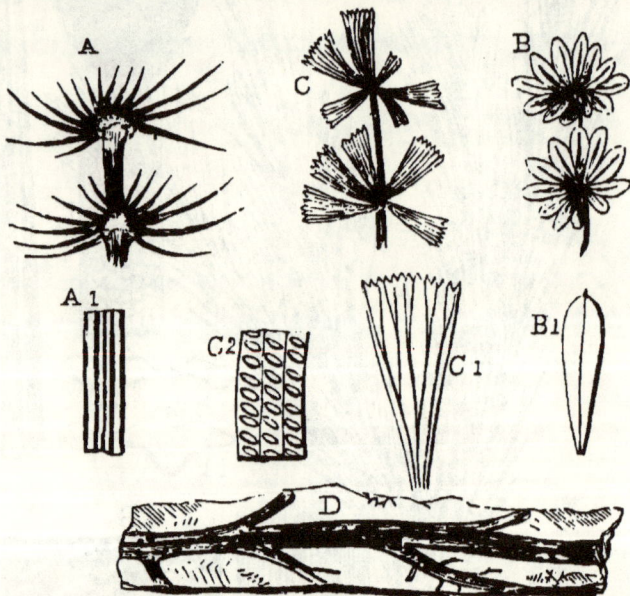

Fɪɢ. 45.—*Asterophyllites*, *Sphenophyllum*, and *Annularia*. ᴀ, *Astero-*
phyllites trinerne. ᴀ¹, Leaf enlarged. ʙ, *Annularia sphenophylloides.*
ʙ¹, Leaf enlarged. ᴄ, *Sphenophyllum erosum.* ᴄ¹, Leaflet enlarged.
ᴄ², Scalariform vessel of *Sphenophyllum.* ᴅ, *Pinnularia ramosissima,*
probably a root.

times we have the form of the outer surface of the woody
cylinder, showing longitudinal ribs, nodes, and marks of
the emission of the branchlets. Sometimes we have the
outer surface of the plant covered with a smooth bark
showing flat ribs, or almost smooth, and having at the
nodes regular articulations with the bases of the verticil-

late branchlets, or on the lower part of the stem the marks of the attachment of the roots.   The Calamites grew in dense clumps, budding off from one another, sometimes at different levels, as the mud or sand accumulated about their stems, and in some species there were creeping rhizomata or root-stocks (Figs. 46 to 49).

But all Calamites were not alike in structure.   In a recent paper *

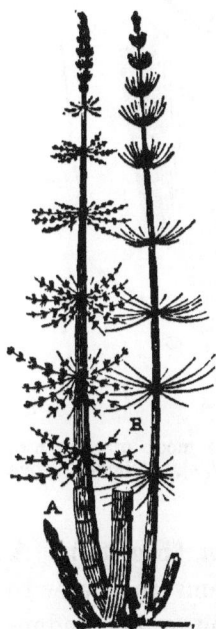

FIG. 46. — *Calamites.*
A, *C. Suckorii.* B,
C. Cistii. (From
" Acadian Geolo-
gy.")

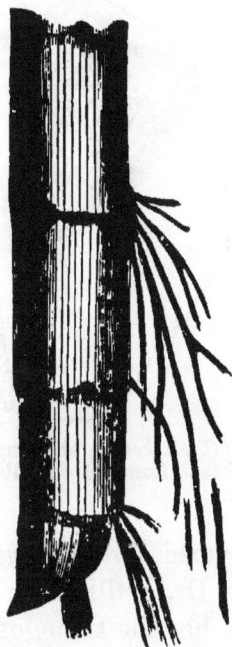

FIG. 47. — Erect *Cala-
mites,* with roots at-
tached (Nova Sco-
tia).

FIG. 48.—Node of *C.
Cistii,* with long
leaves (Nova Sco-
tia).

Dr. Williamson describes three distinct structural types. What he regards as typical *Calamites* has in its woody zone wedges of barred vessels, with thick bands of cellular tissue separating them.   A second type, which

---

* " Memoirs of the Philosophical Society," Manchester, 1886–'87.

he refers to *Calamopitus*, has woody bundles composed of reticulated or multiporous fibres, with their porous sides parallel to the medullary rays, which are better developed than in the previous form. The intervening cellular masses are composed of elongated cells. This is a decided advance in structure, and is of the type of those forms having the most woody and largest stems,

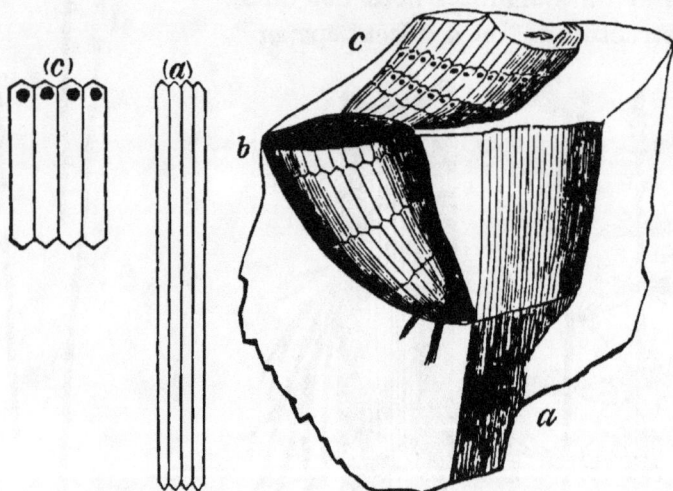

Fig. 49.—Erect *Calamites* (*C. Suckovii*), showing the mode of growth of new stems (*b*), and different forms of the ribs (*a*, *c*). (Pictou, Nova Scotia.) Half natural size.

which Brongniart named *Calamodendron* (Fig. 50). A third form, to which Dr. Williamson seems to prefer to assign this last name, has the tissue of the woody wedges barred, as in the first, but the medullary rays are better developed than in the second. In this third form the intermediate tissue, or primary medullary rays, is truly fibrous, and with secondary medullary rays traversing it. My own observations lead me to infer that there was a fourth type of calamitean stem, less endowed with woody matter, and having a larger fistulous or cellular cavity than any of those described by Dr. Williamson.

There is every reason to believe that all these various

and complicated stems belonged to higher and nobler types of mare's-tails than those of the modern world, and that their fructification was equisetaceous and of the form known as *Calamostachys.*.

We have already seen that noble tree-ferns existed in the Erian period, and these were continued, and their number and variety greatly extended, in the Carboniferous. In regard to the structure of their stems, and the method of supporting these by aërial roots, the tree-ferns of all ages have been nearly alike, and the form and structure of the leaves, except in some comparatively rare and exceptional types, has also been much the same. Any ordinary observer examining a collection of coal-formation ferns recognises at once their kinship to the familiar brackens of our own time. Their fructification is, unfortunately, rarely preserved, so that we are not able, in the case of many species, to speak confidently of

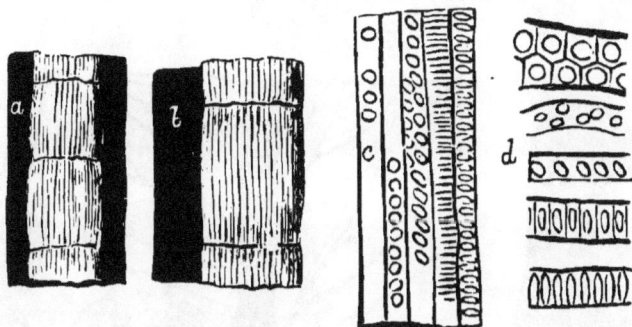

Fig. 50.—Stems of *Calamodendron* and tissues magnified (Nova Scotia). *a*, *b*, Casts of axis in sandstone, with woody envelope (reduced). *c*, *d*, Woody tissue (highly magnified).

their affinities with modern forms ; but the knowledge of this subject has been constantly extending, and a sufficient amount of information has been obtained to enable us to say something as to their probable relationships. (Figs. 51 to 55.)

The families into which modern ferns are divided are, it must be confessed, somewhat artificial, and in the case of fossil ferns, in which the fructification is for the most part wanting, it is still more so, depending in great part on the form and venation of the divisions of the fronds.

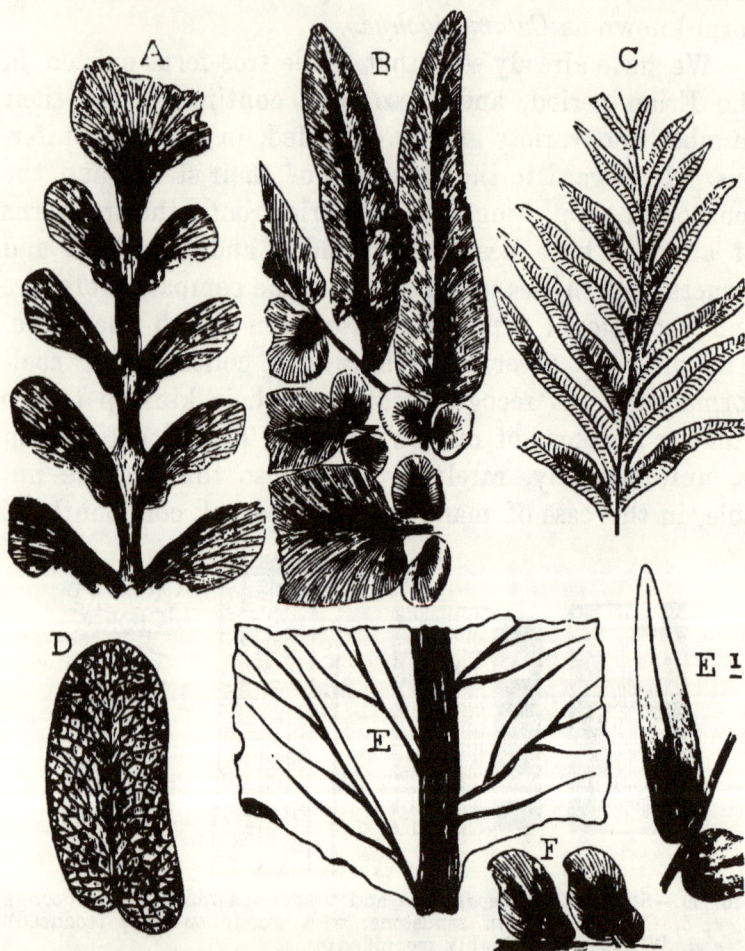

FIG. 51.—Group of coal-formation ferns. A, *Odontopteris subcuneata* (Bunbury). B, *Neuropteris cordata* (Brongniart). C, *Alethopteris lonchitica* (Brongniart). D, *Dictyopteris obliqua* (Bunbury). E, *Phyllopteris antiqua* (Dawson), magnified; E¹, Natural size. F, *Neuropteris cyclopteroides* (Dawson).

Of about eight families into which modern ferns are divided, seven are found in a fossil state, and of these, four at least, the *Cyathaceæ*, the *Ophioglosseæ*, the *Hy-*

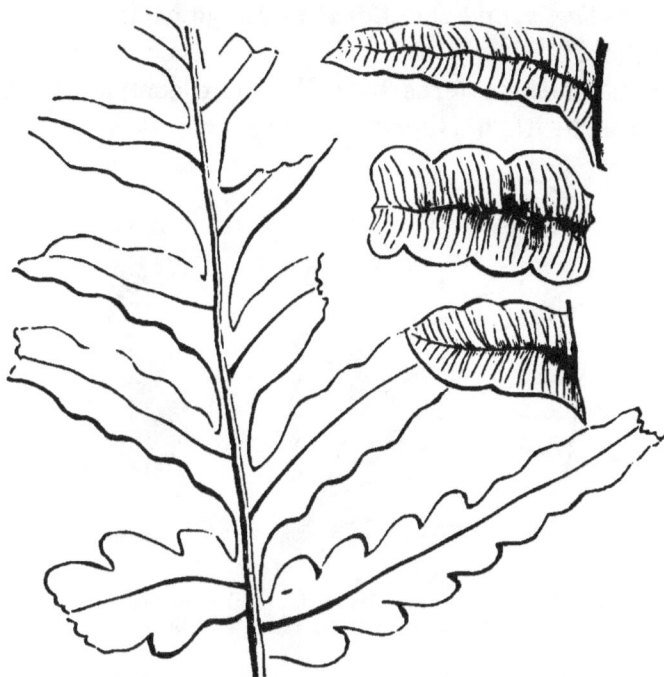

FIG. 52.—*Alethopteris grandis* (Dawson). Middle coal-formation of Nova Scotia.

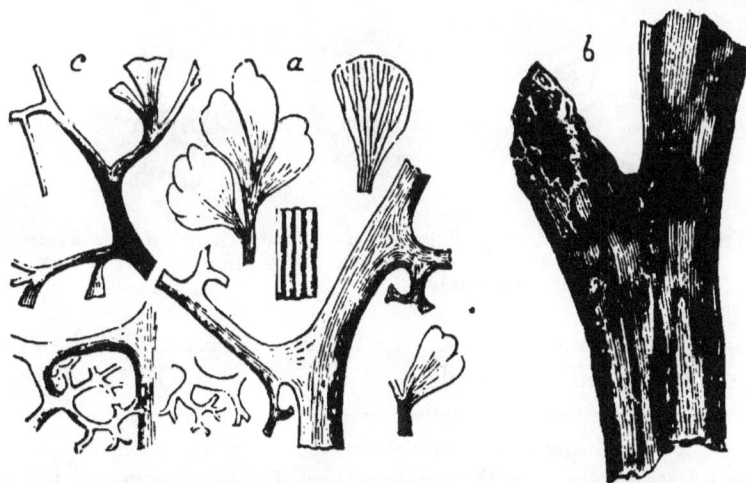

FIG. 53.—*Cyclopteris (Aneimites) Acadica* (Dawson), a tree-fern of the Lower Carboniferous. *a*, Pinnules. *b*, Fragment of petiole. *c*, Remains of fertile pinnules.

menophyllaceæ, and the *Marattiaceæ,* go back to the coal-formation.*

Some of these ferns have the more complex kind of spore-case, with a jointed, elastic ring. It is to be ob-

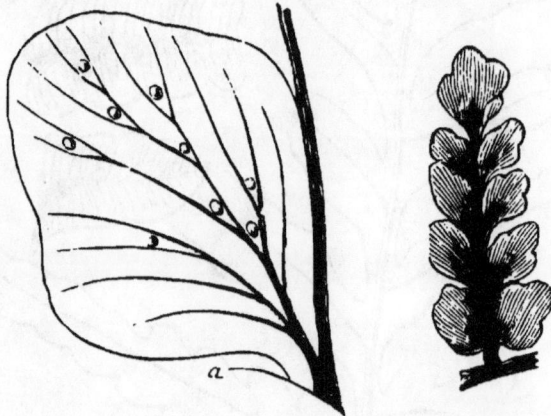

FIG. 54.—*Sphenopteris latior,* Dawson. Coal-formation. *a,* Pinnule magnified, with traces of fructification.

served, however, that those forms which have a simple spore-case, either netted or membranous, and without annulus, are most common in the Devonian and lowest

FIG. 55.—Fructification of Palæozoic ferns. *a,* Thecæ of *Archæopteris* (Erian). *b,* Theca of *Senftenbergia* (Carboniferous). *c,* Thecæ of *Asterotheca* (Carboniferous).

Carboniferous. Some of the forms in these old rocks are somewhat difficult to place in the system. Of these, the

---

* Mr. R. Kidston has recently described very interesting forms of fern fructification from the coal-formation of Great Britain, and much has been done by European palæobotanists, and also by Lesquereux and Fontaine in America.

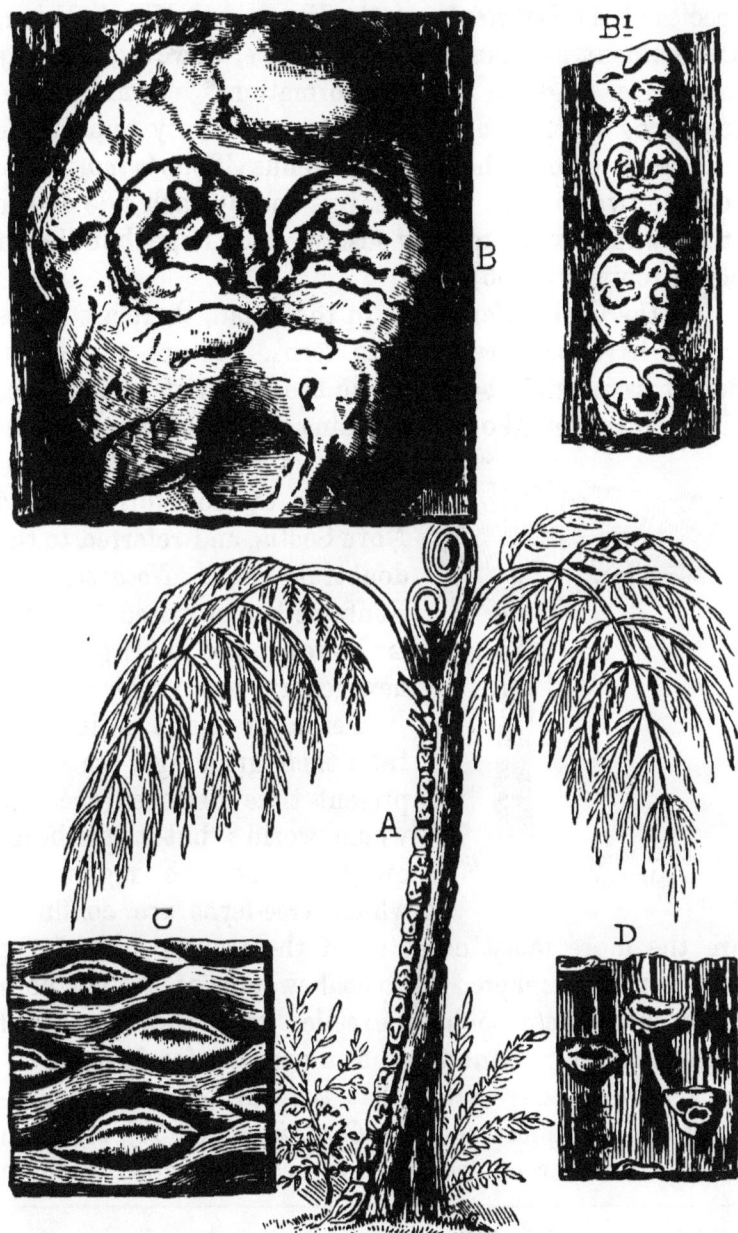

FIG. 56.—Tree-ferns of the Carboniferous. A, *Megaphyton magnificum*, Dawson, restored. B, Leaf-scar of the same, two thirds natural size. B¹, Row of leaf-scars, reduced. C, *Palæopteris Hartii*, scars half natural size. D, *Palæopteris Acadica*, scars half natural size.

species of *Archæopteris*, of the Upper and Middle Erian,
are eminent as examples. This type, however, scarcely
extends as high as the coal-formation.* Some of the
tree-ferns of the Carboniferous present very remarkable
features. One of these, of the genus *Megaphyton*, seems
to have two rows of great leaves, one at each side of the
stem, which was probably sustained by large bundles of
aërial roots (Fig. 56).

In the Carboniferous, as in the Erian, there are leaves
which have been referred to ferns, but are subject to
doubt, as possibly belonging to broad-leaved taxine trees
allied to the gingko-tree of China. One of these, represented in Fig. 57, has been
found in the coal-formation of
Nova Scotia, and referred to the
doubtful genus *Noeggerathia*.
Fontaine has proposed for similar leaves found in Virginia the
new generic name *Saportea*.

FIG. 57.—*Noeggerathia dispar*
(half natural size).

Ferns, as might be inferred
from their great age, are at the
present time dispersed over the
whole world; but their headquarters, and the regions to
which tree-ferns are confined,
are the more moist climates of the tropics and of the
southern hemisphere. The coal-swamps of the northern
hemisphere seem to have excelled even these favoured
regions of the present world as a paradise for ferns.

I have already stated that the Carboniferous constitutes the headquarters of the *Cordaites* (Fig. 58), of which
a large number of species have been described, both in

---

* The pretty little ferns of the genus *Botrychium* (moonwort), so
common in American and European woods, seem to be their nearest modern allies.

Europe and America. We sometimes, though rarely, find their stems showing structure. In this case we have a large cellular pith, often divided by horizontal partitions into flat chambers, and constituting the objects which, when detached, are called *Sternbergiæ* (Fig. 62). These Sternbergia piths, however, occur in true conifers as well, as they do in the modern world in some trees, like our common butternut, of higher type ; and I showed many years ago that the Sternbergia type may be detected in the young twigs of the balsam - fir (*Abies balsamifera*). The pith was surrounded by a ring of scalariform or barred tissue, often of considerable thickness, and in young stems so important as to have suggested lycopodiaceous affinities. But as the stem grew in size, a regular ring of woody wedges, with tissue having rounded or hexagonal pores or discs,

Fig. 58.—*Cordaites* (*Dorycordaites*), Grand' Eury, reduced.

like those of pines, was developed. Outside this was a bark, often apparently of some thickness. This structure in many important points resembles that of cycads, and also approaches to the structure of Sigillaria, while in its more highly developed forms it approximates to that of the conifers.

On the stems so constructed were placed long and
often broad many-nerved leaves, with rows of stomata or
breathing-pores, and attached by somewhat broad bases
to the stem and branches. The fruit consisted of racemes,
or clusters of nutlets, which seem to have been provided

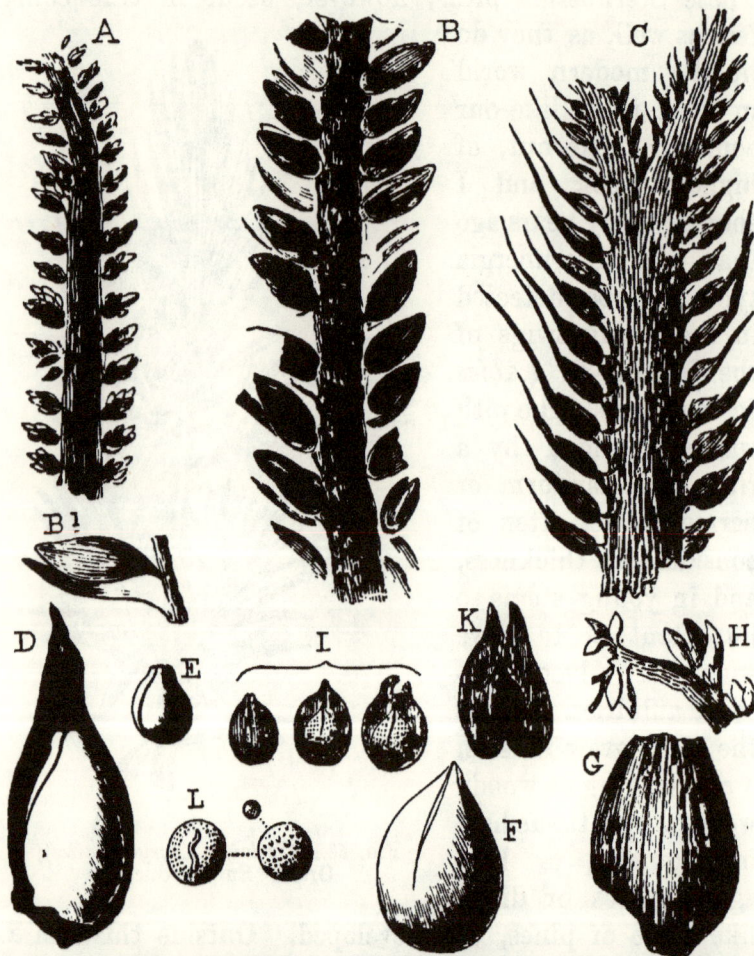

FIG. 59.—Fruits of *Cordaites* and Taxine Conifers (coal-formation, Nova
Scotia.) A, *Antholithes squamosus* (two thirds). B, *A. rhabdocarpi*
(two thirds). B¹, Carpel restored. C, *A. spinosus* (natural size).
D, *Trigonocarpum intermedium*. E, *T. Nœggerathii*. F, *T. avella-
num*. G, *Rhabdocarpus insignis*, reduced. H, *Antholithes pygmæus*.
I, *Cardiocarpum fluitans*. K, *Cardiocarpum bisectum*. L, *Sporangites
papillata*, lycopodiaceous macrospores (natural size and magnified).

with broad lateral wings for flotation in the air, or in some cases with a pulpy envelope, which flattens into a film. There seem to have been structures of both these kinds, though in the state of preservation of these curious seeds it is extremely difficult to distinguish them. In the first case they must have been intended for dissemination by the wind, like the seeds of spruces. In the latter case they may have been disseminated like the fruits of taxine trees by the agency of animals, though what these were it would be difficult to guess. These trees had very great reproductive power, since they produced numerous seeds, not singly or a few together, as in modern yews, but in long spikes or catkins bearing many seeds (Fig. 59).

It is to be observed that the Cordaites, or the *Cordaitinæ*, as they have been called, as a family,* constitute another of those intermediate groups with which we have already become familiar. On the one hand they approach closely to the broader-leaved yews like Gingko, Phyllocladus, and Podocarpus, and, on the other hand, they have affinities with Cycadaceæ, and even with Sigillariæ. They were beautiful and symmetrical trees, adding something to the variety of the rather monotonous Palæozoic forests. They contributed also somewhat to the accumulation of coal. I have found that some thin beds are almost entirely composed of their leaves, and the tissues of their wood are not infrequent in the mineral charcoal of the larger coal-seams. There is no evidence that their roots were of the stigmaroid type, though they evidently grew in the same swampy flats with the Sigillariæ and Calamites.

It may, perhaps, be well to say here that I believe there was a considerably wide range of organisation in the Cordaitinæ as well as in the Calamites and Sigillariæ, and that it will eventually be found that there were three lines

---

* Engler ; Cordaitées of Renault.

of connection between the higher cryptogams and the phænogams, one leading from the lycopods by the Sigillariæ, another leading by the Cordaites, and the third leading from the Equisetums by the Calamites. Still further back the characters afterward separated in the club-mosses, mare's-tails, and ferns, were united in the Rhizocarps, or, as some now, but I think somewhat unreasonably, prefer to call them, the "heterosporous Filicinæ." In the more modern world, all the connecting links have become extinct and the phænogams stand widely separated from the higher cryptogams. I do not make these remarks in a Darwinian sense, but merely to state what appear to be the lines of natural affinity and the links wanting to give unity to the system of nature.

Of all the trees of the modern world, none are perhaps so widely distributed as the pines and their allies. On mountain-tops and within the Arctic zone, the last trees that can struggle against the unfavourable conditions of existence are the spruces and firs, and in the warm and moist islands of the tropics they seem equally at home with the tree-ferns and the palms. We have already seen that they are a very ancient family, and in the sandstones of the coal-formation their great trunks are frequently found, infiltrated with calcareous or silicious matter, and still retaining their structure in the greatest perfection (Fig. 60). So far as we know, the foliage of some of them which constitutes the genera *Walchia* and *Araucarites* of some authors (Figs. 60, 63) was not dissimilar from that of modern yews and spruces, though there is reason to believe that some others had broad, fern-like leaves like those of the gingko. None of them, so far as yet certainly known, were cone-bearing trees, their fruit having probably been similar to that of the yews (Fig. 61). The minute structures of their stems are nearer to those of the conifers of the islands of the southern hemisphere than to that of those in our northern climes—a cor-

FIG. 60.—Coniferous wood and foliage (Carboniferous). A, *Araucarites gracilis*, reduced. B, *Dadoxylon Acadianum* (radial), 90 diams.; B¹ (tangential), 90 diams.; B², cell showing areolation, 250 diams. C, *Dadoxylon materiarium* (radial), 90 diams.; C¹ (tangential), 90 diams.; C², cell showing areolation, 250 diams. D, *Dadoxylon antiquius* (radial), 90 diams.; D¹ (tangential), 90 diams.; D², cell showing areolation, 250 diams.

relation, no doubt, to the equable climate of the period.
There is not much evidence that they grew with the Sigillariæ in the true coal-swamps, though some specimens
have been found in this association.   It is more likely
that they were in the main inland and upland trees, and

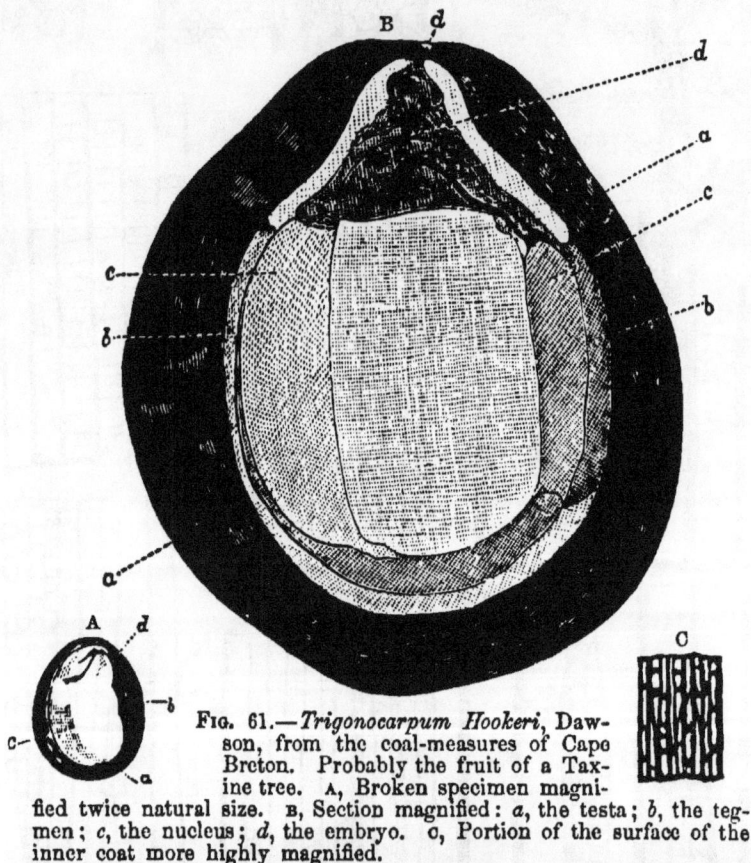

Fig. 61.—*Trigonocarpum Hookeri*, Dawson, from the coal-measures of Cape
Breton. Probably the fruit of a Taxine tree. A, Broken specimen magni-
fied twice natural size. B, Section magnified: *a*, the testa; *b*, the tegmen; *c*, the nucleus; *d*, the embryo. C, Portion of the surface of the
inner coat more highly magnified.

that in consequence they are mostly known to us by
drifted trunks borne by river inundations into the seas
and estuaries.

A remarkable fact in connection with them, and showing also the manner in which the most durable vegetable
structures may perish by decay, is that, like the Cordaites,
they had large piths with transverse partitions, a struct-

ure which, as I have already mentioned, appears on a
minute scale in the twigs of the fir-tree, and that sometimes casts of these piths in sandstone appear in a separate
form, constituting what have been named *Sternbergiæ* or
Artisiæ. As Renault well remarks with reference to
Cordaites, the existence of this chambered form of pith
implies rapid elongation of the stem, so that the Cordaites
and conifers of the coal-formation were probably quickly
growing trees (Fig. 62).

The same general statements may be made as to the
coal-vegetation as in relation to that of the Erian. In

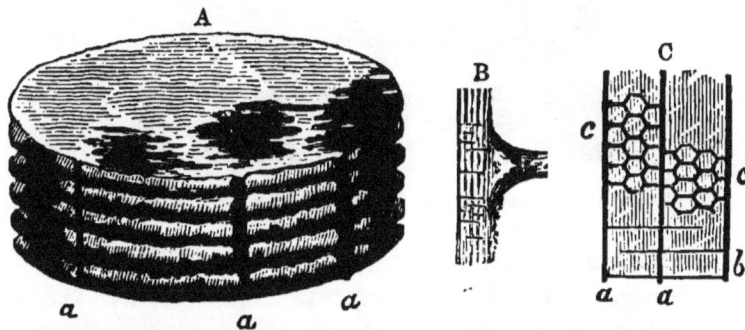

Fig. 62.—*Sternbergia* pith of *Dadoxylon*. A, Specimen (natural size),
showing remains of wood at *a, a*. B, Junction of wood and pith, magnified. c, Cells of the wood of do., *a, a*; *b*, medullary ray; *c*, areolation.

the coal period we have found none of the higher exogens, and there are only obscure and uncertain indications of the presence of endogens, which we may reserve
for a future chapter ; but gymnosperms abound and are
highly characteristic. On the other hand, we have no
mosses or lichens, and very few Algæ, but a great number of ferns and Lycopodiaceæ or club-mosses (Fig. 63).
Thus, the coal-formation period is botanically a meeting-
place of the lower phænogams and the higher cryptogams,
and presents many forms which, when imperfectly known,
have puzzled botanists in regard to their position in one
or other series. In the present world, the flora most akin

to that of the coal period is that of warm, temperate regions in the southern hemisphere.   It is not properly a tropical flora, nor is it the flora of a cold region, but rather indicative of a moist and equable climate.   Still,

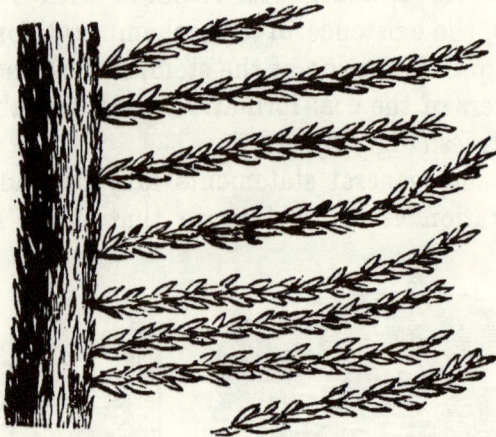

FIG. 63.—*Walchia imbricatula*, S. N., Permian, Prince Edward Island.

we must bear in mind that we may often be mistaken in reasoning as to the temperature required by extinct species of plants, differing from those now in existence. Further, we must not assume that the climatal conditions of the northern hemisphere were in the coal period at all similar to those which now prevail.   As Sir Charles Lyell has shown, a less amount of land in the higher latitudes would greatly modify climates, and there is every reason to believe that in the coal period there was less land than now.   Further, it has been shown by Tyndall that a very small additional amount of carbonic acid in the atmosphere would, by obstructing the radiation of heat from the earth, produce almost the effect of a glass roof or conservatory, extending over the whole world.   Again, there is much in the structure of the leaves of the coal-plants, as well as in the vast amount of carbon which they accumulated in the form of coal, and the characteristics of the animal life of the period, to indicate, on independent grounds, that the carboniferous atmosphere differed from that of the present world in this way, or in the presence of more carbonic acid—a substance now existing in the very minute proportion of one thousandth of the whole—a quantity adapted to the present requirements of vegetable and animal life, but probably not to those of the coal period.

Thus, if we inquire as to any analogous distribution of plants in the modern world, we find this only in the warmer insular climates of the southern hemisphere, where ferns, lycopods, and pines appear under forms somewhat akin to those of the Carboniferous, but mixed with other types, some of which are modern, others allied to those of the next succeeding geological ages of the Mesozoic and Tertiary ; and under these periods it will be more convenient to make comparisons.

The readers of recent English popular works on geology will have observed the statement reiterated that a large proportion of the material of the great beds of bituminous coal is composed of the spore-cases of lycopodiaceous plants—a statement quite contrary to that resulting from my microscopical examinations of the coal of more than eighty coal-beds in Nova Scotia and Cape Breton, as stated in "Acadian Geology" (page 463), and more fully in my memoir of 1858 on the "Structures in Coal," * and that of 1866, on the "Conditions of Accumulation of Coal." † The reason of this mistake is, that an eminent English naturalist, happening to find in certain specimens of English coal a great quantity of remains of spores and spore-cases, though even in his specimens they constitute only a small portion of the mass, and being apparently unacquainted with what others had done in this field, wrote a popular article for the "Contemporary Review," in which he extended an isolated and

* " Journal of the Geological Society," vol. xv.   † Ibid., vol. xxii.

exceptional fact to all coals, and placed this supposed
origin of coal in a light so brilliant and attractive that he
has been followed by many recent writers.   The fact is,
as stated in "Acadian Geology," that trunks of *Sigillariæ*
and similar trees constitute a great part of the denser
portion of the coal, and that the cortical tissues of these
rather than the wood remain as coal.   But cortical or
epidermal tissues in general, whether those of spore-cases
or other parts of plants, are those which from their resistance to water-soakage and to decay, and from their
highly carbonaceous character, are best suited to the production of coal.   In point of fact, spore-cases, though
often abundantly present, constitute only an infinitesimal
part of the matter of the great coal-beds.   In an article
in "The American Journal of Science," which appeared
shortly after that above referred to, I endeavoured to correct this error, though apparently without effect in so far
as the majority of British geological writers are concerned.   From this article I have taken with little change
the following passages, as it is of importance in theoretical
geology that such mistakes, involving as they do the
whole theory of coal accumulation, should not continue
to pass current.   The early part of the paper is occupied
with facts as to the occurrence of spores and spore-cases as
partial ingredients in coal.   Its conclusions are as follows :

It is not improbable that sporangites, or bodies resembling them, may be found in most coals; but it
is most likely that their occurrence is accidental rather
than essential to coal accumulation, and that they are
more likely to have been abundant in shales and cannel
coals, deposited in ponds or in shallow waters in the vicinity of lycopodiaceous forests, than in the swampy
or peaty deposits which constitute the ordinary coals.
It is to be observed, however, that the conspicuous appearance which these bodies, and also the strips and
fragments of epidermal tissue, which resemble them in

texture, present in slices of coal, may incline an observer, not having large experience in the examination of coals, to overrate their importance ; and this I think has been done by most microscopists, especially those who have confined their attention to slices prepared by the lapidary. One must also bear in mind the danger arising from mistaking concretionary accumulations of bituminous matter for sporangia. In sections of the bituminous shales accompanying the Devonian coal above mentioned, there are many rounded yellow spots, which on examination prove to be the spaces in the epidermis of *Psilophyton* through which the vessels passing to the leaves were emitted. To these considerations I would add the following, condensed from the paper above referred to (p. 139), in which the whole question of the origin of coal is fully discussed :*

1. The mineral charcoal or 'mother coal' is obviously woody tissue and fibres of bark, the structure of the varieties of which, and the plants to which it probably belongs, I have discussed in the paper above mentioned.

2. The coarser layers of coal show under the microscope a confused mass of fragments of vegetable matter belonging to various descriptions of plants, and including, but not usually in large quantities, sporangites.

3. The more brilliant layers of the coal are seen, when separated by thin laminæ of clay, to have on their surfaces the markings of *Sigillariæ* and other trees, of which they evidently represent flattened specimens, or rather the bark of such specimens. Under the microscope, when their structures are preserved, these layers show cortical tissues more abundantly than any others.

4. Some thin layers of coal consist mainly of flattened layers of leaves of *Cordaites* or *Pychnophyllum*.

5. The *Stigmaria* underclays and the stumps of

---

* See also "Acadian Geology," 2d ed., pp. 138, 461, 493.

Sigillaria in the coal-roofs equally testify to the accumulation of coal by the growth of successive forests, more especially of *Sigillariæ*.   There is, on the other hand, no necessary connection of sporangite-beds with Stigmarian soils.   Such beds are more likely to be accumulated in water, and consequently to constitute bituminous shales and cannels.

6. *Lepidodendron* and its allies, to which the spore-cases in question appear to belong, are evidently much less important to coal accumulation than *Sigillaria*, which cannot be affirmed to have produced spore-cases similar to those in question, even though the observation of Goldenberg as to their fruit can be relied on ; the accuracy of which, however, I am inclined to doubt.

On the whole, then, while giving due credit to those who have advocated the spore-theory of coal, for directing attention to this curious and no doubt important constituent of mineral fuel, and admitting that I may possibly have given too little attention to it, I must maintain that sporangite-beds are exceptional among coals, and that cortical and woody matters are the most abundant ingredients in all the ordinary kinds; and to this I cannot think that the coals of England constitute an exception.

It is to be observed, in conclusion, that the spore-cases of plants, in their indestructibility and richly carbonaceous character, only partake of qualities common to most suberous and epidermal matters, as I have explained in the publications already referred to.   Such epidermal and cortical substances are extremely rich in carbon and hydrogen, in this resembling bituminous coal.   They are also very little liable to decay, and they resist more than other vegetable matters aqueous infiltration—properties which have caused them to remain unchanged, and to continue free from mineral additions more than other vegetable tissues.   These qualities are well seen in the bark of our American white birch.   It is no wonder that materials of this kind should constitute considerable portions of such vegetable accumulations as the beds of coal, and that when present in large proportion they should afford richly bituminous beds. All this agrees with the fact, apparent on examination of the common coal, that the greater number of its purest layers consist of the flattened bark of *Sigillariæ* and similar trees, just as any single flattened trunk embedded in shale becomes a layer of pure coal. It also agrees with the fact that other layers of coal, and also the cannels and earthy bitumens, appear under the microscope to consist of finely comminuted particles, principally of epidermal tissues, not only from the fruits and spore-cases of plants, but also from their leaves and stems. These considerations impress us, just as much as the abundance of spore-cases, with the immense amount of the vegetable matter which has perished during the accumulation of coal, in comparison with that which has been preserved.

I am indebted to Dr. T. Sterry Hunt for the following very valuable information, which at once places in a clear and precise light the chemical relations of epidermal tissue and spores with coal. Dr. Hunt says : "The outer bark of the cork-tree, and the cuticle of many if not all other plants, consists of a highly carbonaceous matter, to which the name of *suberin* has been given. The spores of *Lycopodium* also approach to this substance in composition, as will be seen by the following, one of two analyses by Duconi,* along with which I give the theoretical composition of pure cellulose or woody fibre, according to Payen and Mitscherlich, and an analysis of the suberin of cork, from *Quercus suber*, from which the ash and 2·5 per cent of cellulose have been deducted.†

---

* Liebig and Kopp, "Jahresbuch," 1847–'48.
† Gmelin, "Handbook," xv., 145.

	Cellulose.	Cork.	Lycopodium.
Carbon	44·44	65·73	64·80
Hydrogen	6·17	8·33	8·73
Nitrogen	....	1·50	6·18
Oxygen	49·39	24·44	20·29
Total	100·00	100·00	100·00

"This difference is not less striking when we reduce the above centesimal analyses to correspond with the formula of cellulose, $C_{24}H_{20}O_{20}$, and represent cork and *Lycopodium* as containing twenty-four equivalents of carbon. For comparison I give the composition of specimens of peat, brown coal, lignite, and bituminous coal :[*]

Cellulose .... $C_{24}H_{20}O_{20}$
Cork .... $C_{24}H_{18\frac{1}{2}}O_{6\frac{7}{10}}$
Lycopodium .... $C_{24}H_{19\frac{4}{10}}NO_{6\frac{6}{10}}$
Peat (Vaux) .... $C_{24}H_{14\frac{7}{10}}O_{10}$
Brown coal (Schröther) .... $C_{24}H_{14\frac{3}{10}}O_{10\frac{6}{10}}$
Lignite (Vaux) .... $C_{24}H_{11\frac{3}{10}}O_{6\frac{4}{10}}$
Bituminous coal (Regnault) ..... $C_{24}H_{10}O_{3\frac{2}{10}}$

"It will be seen from this comparison that, in ultimate composition, cork and *Lycopodium* are nearer to lignite than to woody fibre, and may be converted into coal with far less loss of carbon and hydrogen than the latter. They in fact approach closer in composition to resins and fats than to wood, and, moreover, like those substances repel water, with which they are not easily moistened, and thus are able to resist those atmospheric influences which effect the decay of woody tissue."

I would add to this only one further consideration. The nitrogen present in the *Lycopodium* spores, no doubt, belongs to the protoplasm contained in them, a substance which would soon perish by decay ; and subtracting this, the cell-walls of the spores and the walls of the spore-

---

[*] "Canadian Naturalist," vi., 253.

cases would be most suitable material for the production of bituminous coal. But this suitableness they share with the epidermal tissue of the scales of strobiles, and of the stems and leaves of ferns and lycopods, and, above all, with the thick, corky envelope of the stems of *Sigillariæ* and similar trees, which, as I have elsewhere shown,[*] from its condition in the prostrate and erect trunks contained in the beds associated with coal, must have been highly carbonaceous and extremely enduring and impermeable to water. In short, if, instead of "spore-cases," we read "epidermal tissues in general, including spore-cases," all that has been affirmed regarding the latter will be strictly and literally true, and in accordance with the chemical composition, microscopical characters, and mode of occurrence of coal. It will also be in accordance with the following statement, from my paper on the "Structures in Coal," published in 1859 :

"A single trunk of *Sigillaria* in an erect forest presents an epitome of a coal-seam. Its roots represent the *Stigmaria* underclay; its bark the compact coal; its woody axis the mineral charcoal; its fallen leaves (and fruits), with remains of herbaceous plants growing in its shade, mixed with a little earthy matter, the layers of coarse coal. The condition of the durable outer bark of erect trees concurs with the chemical theory of coal, in showing the especial suitableness of this kind of tissue for the production of the purer compact coals. It is also probable that the comparative impermeability of the bark to mineral infiltration is of importance in this respect, enabling this material to remain unaffected by causes which have filled those layers, consisting of herbaceous materials and decayed wood, with pyrites and other mineral substances."

---

[*] "Vegetable Structures in Coal," "Journal of Geological Society," xv., 626. "Conditions of Accumulation of Coal," *ibid.*, xxii., 95. "Acadian Geology," 197, 464.

We need not go far in search of the uses of the coal
vegetation, when we consider the fact that the greatest
civilised nations are dependent on it for their fuel. Without the coal of the Carboniferous period and the iron-ore
which is one of the secondary consequences of coal accumulation, just as bog-ores of iron occur in the subsoils
of modern peats, it would have been impossible either to
sustain great nations in comfort in the colder climates of
the northern hemisphere or to carry on our arts and
manufactures. The coal-formation yields to Great Brit-
ian alone about one hundred and sixty million tons of
coal annually, and the miners of the United States extract mainly from the same formation nearly a hundred
million tons, while the British colonies and dependen-
cies produce about five million tons ; and it is a remarkable fact that it is to the English race that the
greatest supply of this buried power and heat and light
has been given.

The great forests of the coal period, while purifying
the atmosphere of its excess of unwholesome carbonic
acid, were storing up the light and heat of Palæozoic
summers in a form in which they could be recovered in our
human age, so that, independently of their uses to the
animals which were their contemporaries, they are indis-
pensable to the existence of civilised man.

Nor can we hope soon to be able to dispense with the
services of this accumulated store of fuel. The forests
of to-day are altogether insufficient for the supply of our
wants, and though we are beginning to apply water-power
to the production of electricity, and though some promis-
ing plans have been devised for the utilisation of the
direct heat and light of the sun, we are still quite as dependent as any of our predecessors on what has been done
for us in the Palæozoic age.

In the previous pages I have said little respecting the
physical geography of the Carboniferous age ; but, as may

be inferred from the vegetation, this in the northern hemisphere presented a greater expanse of swampy flats little elevated above the sea than we find in any other period. As to the southern hemisphere, less is known, but the conditions of vegetation would seem to have been essentially the same.

Taking the southern hemisphere as a whole, I have not seen any evidence of a Lower Devonian or Upper Silurian flora; but in South Africa and Australia there are remains of Upper Devonian or Lower Carboniferous plants. These were succeeded by a remarkable Upper Carboniferous or Permian group, which spread itself all over India, Australia, and South Africa,* and contains some forms (*Vertebraria, Phyllotheca, Glossopteris, &c.*) not found in rocks of similar age in the northern hemisphere, so that, if the age of these beds has been correctly determined, the southern hemisphere was in advance in relation to some genera of plants. This, however, is to be expected when we consider that the Triassic and Jurassic flora of the north contains or consists of intruders from more southern sites. These beds are succeeded in India by others holding cycads, &c., of Upper Jurassic or Lower Cretaceous types (Rajmahal and Jabalpur groups).

Blanford has shown that there is a very great similarity in this series all over the Australian and Indian region.† Hartt and Darby have in like manner distinguished Devonian and Carboniferous forms in Brazil akin to those of the northern hemisphere. Thus the southern hemisphere would seem to have kept pace with the northern, and according to Blanford there is evidence there of cold conditions in the Permian, separating the Palæozoic

---

* Wyley, " Journal Geol. Society," vol. xxiii., p. 172 ; Daintree, *ibid.*, vol. xxviii. ; also Clarke and McCoy.

† " Journal Geol. Society," vol. xxxi.

flora from that of the Mesozoic, in the same manner that Ramsay has supposed a similar period of cold to have done north of the equator. This would imply a very great change of climate, since we have evidence of the extension of the Lower Carboniferous flora at least as far north as Spitzbergen. The upper coal-formation we cannot, however, trace nearly so far north; so that a gradual refrigeration may have been going on before the Permian. Thus in both hemispheres there was a general similarity in the later Palæozoic flora, and perhaps similar conditions leading to its extinction and to its replacement by that to be described in the next chapter.

NOTES TO CHAPTER IV.

I. CHARACTERS AND CLASSIFICATION OF PALÆOZOIC PLANTS.

IN the space available in this work it would be impossible to enter fully into the classification of Palæozoic plants; but it may be well to notice some important points for the guidance of those who may desire to collect specimens; more especially as much uncertainty exists as to affinities and very contradictory statements are made. The statements below may be regarded as the results of actual observation and of the study of specimens *in situ* in the rocks, as well as in the cabinet and under the microscope.

GYMNOSPERMEÆ.

Family CONIFERÆ; *Genus* DADOXYLON, Endlicher; ARAUCARITES, Goeppert; ARAUCARIOXYLON, Kraus.

The trunks of this genus occur from the Middle Devonian to the Permian inclusive, as drift-logs calcified, silicified, or pyritised. The only foliage associated with them is of the type of *Walchia* and *Araucarites*—viz., slender branches with numerous small spiral acicular leaves. Two of the coal-formation species, *D. materiarum* and another, had foliage of this type. That of the others is unknown. They are all distinct from the wood of *Cordaites*, for which see under that genus.

The following are North American species:

Trunks.

Dadoxylon Ouangondianum, Dn .. M. Erian..... ...Report, 1871.*
D. Halli, Dn................... " .... "
D. Newberryi, Dn............... " .... "
D. Clarkii, Dn. (Cordæoxylon ?)... " ....Report, 1882.
D. Acadianum, Dn..............Coal - formation Acadian Geol-
                                and millstone ogy.
                                grit.
D. Materiarum, Dn ....Do. and Permo- "
                                Carb.
D. (Palæoxylon) antiquius, Dn ...L. Carboniferous. "
D. annulatum, Dn ....Coal-formation. "
Ormoxylon Erianum, Dn ....Erian............Report, 1871.

Foliage.

Araucarites gracilis, Dn.........N. Coal-formation "
                                and Permian.

Walchia robusta, Dn............Permian.   ⎧ Report on
W. imbricatula, Dn............. "   ⎨ Prince Ed-
                                              ⎩ ward Island.

All of the above can be vouched for as good species based upon microscopic examination of a very large number of trunks from different parts of North America. The three Erian species of *Dadoxylon* and *D. antiquius* from the Lower Carboniferous have two or more rows of cells in the medullary rays. The last named has several rows, and is a true *Palæoxylon* allied to *D. Withami* of Great Britain. *D. materiarium* is specially characteristic of the upper coal-formation and Permian, and to it must belong one or both of the species of foliage indicated above. *D. Clarkii* has very short, simple medullary rays of only a few cells superimposed, and has an inner cylinder of scalariform vessels, approaching in these points to *Cordaites. Ormoxylon* has a very peculiar articulated pith and simple medullary rays.

Witham in 1833 described several Carboniferous species of pine-wood, under the generic name *Pinites*, separating under the name *Pitus* species which appeared to have the discs on the cell-walls

---

* "Geological Survey of Canada: Fossil Plants of Erian and Upper Silurian Formations," by J. W. Dawson.

separate and in transverse lines. Witham's name was changed by
Goeppert to *Araucarites*, to indicate the similarity of these woods to
Araucaria, *Pinites* being reserved for trees more closely allied to the
ordinary pines. Endlicher, restricting *Araucarites* to foliage, etc.,
of Araucaria-like trees, gave the name *Dadoxylon* to the wood; and
this, through Unger's "Genera and Species," has gained somewhat
general acceptance. Endlicher also gave the name *Pissadendron* to
the species which Witham had called *Pitus;* but Brongniart proposed the name *Palæoxylon* to include all the species with thick
and complex medullary rays, whatever the arrangement of the discs.
In Schimper's new work Kraus substitutes *Araucarioxylon* for Endlicher's *Dadoxylon*, and includes under *Pissadendron* all the species
placed by Brongniart in *Palæoxylon*.

To understand all this confusion, it may be observed that the
characters available in the determination of Palæozoic coniferous
wood are chiefly the form and arrangement of the wood-cells, the
character of the bordered pores or discs of their walls, and the form
and composition of the medullary rays.

The character on which Witham separated his genus *Pitus* from
Pinites is, as I have ascertained by examination of slices of one of
his original specimens kindly presented to me by Mr. Sanderson, of
Edinburgh, dependent on state of preservation, the imperfectly preserved discs or areolations of the walls of the fibre presenting the
appearance of separate and distinct circles, while in other parts of
the same specimens these discs are seen to be contiguous and to assume hexagonal forms, so that in this respect they do not really
differ from the ordinary species of *Dadoxylon*. The true character
for subdividing those species which are especially characteristic of
the Carboniferous, is the composite structure of the medullary rays,
which are thick and composed of several radial piles of cells placed
side by side. This was the character employed by Brongniart in
separating the genus *Palæoxylon*, though he might with convenience
have retained Witham's name, merely transferring to the genus the
species of Witham's *Pinites* which have complex medullary rays.
The Erian rocks present the greatest variety of types, and *Palæoxylon*
is especially characteristic of the Lower Carboniferous, while species
of *Dadoxylon* with two rows of bordered pores and simple medullary
rays are especially plentiful in the upper coal-formation and Permo-
Carboniferous.

The following table will clearly show the distinctive characters
and relations of the genera in question, as held by the several authors
above referred to:

Wood of Palæozoic Conifers.

Woody fibres.	Medullary rays and pith.	Generic names.	Geological age.
No discs.	One or two series of cells.	*Aporoxylon*, Unger.	Devonian (Erian).
Discs in one series contiguous, or in several series spirally arranged.	Complex, or of two or more series of cells. Pith Sternbergian.	{ *Pitus*, Witham. *Palæoxylon*, Brongniart. *Pissadendron*, Endlicher.	Middle and Lower Carboniferous and Devonian.
	Simple, or of one row of cells. Pith Sternbergian.	{ *Araucarites*, Goeppert *Dadoxylon*, Endlicher. *Araucarioxylon*, Schimper.	Upper Carboniferous and Permian.
	Pith in spherical chambers.	*Ormoxylon*,* Dn.	Devonian.
	Medullary sheath scalariform. Medullary rays frequent, simple, short.	*Dadoxylon* (Cordaoxylon),† Dn.	Devonian.

* Type *O. Erianum*, Dn., "Report on Canadian Plants," 1871.

† Type *D. Clarkii*, Dn., "Report on Canadian Plants," 1882. This may be wood of Cordaites, to which it approaches very closely.

Family CORDAITEÆ, *Genus* CORDAITES, Brongniart.

Trunks marked by transverse scars of attachment of bases of leaves; leaves broad, with many parallel veins, and attached by a broad base; pistillate and staminate catkins of the nature of Antholithes. Fruit winged or pulpy, of the kind known as *Cardiocarpum.* Stem with a Sternbergia pith, usually large, surrounded by a ring of pseudo-scalariform vessels, and with a cylinder usually narrow, of woody wedges, with bordered pores in one or more series, and with simple medullary rays.

From specimens kindly presented to me by Prof. Renault, I have been able to ascertain that the stems of some at least of these plants (Eucordaites) are distinct in structure from all the species of *Dadoxylon*, above mentioned, except *D. Clarkii*, of the Erian. They may be regarded as intermediate between those of conifers and cycads, which is indeed the probable position of these remarkable plants.

Grand Eury has divided the *Cordaites* into sub-genera, as follows:

1. *Eucordaites.*—Leaves spatulate, obovate, elliptical, or lan-

ceolate, sessile, entire, with rounded apices and of leathery consistency. The leaves are from twenty to ninety centimetres in length. The nerves are either equally or unequally strong.

2. *Dorycordaites.*—Leaves lanceolate, with sharp points; nerves numerous, fine, and equal in strength. The leaves attain a length of from forty to fifty centimetres.

3. *Poacordaites.*—Leaves narrow, linear, entire, blunt at the point, with nerves nearly equally strong. The leaves are as much as forty centimetres in length.

To these Renault and Zeiller have added a fourth group, *Scutocordaites.*

Genus STERNBERGIA.

This is merely a provisional genus intended to receive casts of the pith cylinders of various fossil trees. Their special peculiarity is that, as in the modern *Cecropia peltata*, and some species of *Ficus*, the pith consists of transverse dense partitions which, on the elongation of the internodes, become separated from each other, so as to produce a chambered pith cavity, the cast of which shows transverse furrows. The young twigs of the modern *Abies balsamifera* present a similar structure on a minute scale. I have ascertained and described such pith-cylinders in large stems of *Dadoxylon Ouangondianum*, and *D. materiarium.* They occur also in the stems of *Cordaites* and probably of *Sigillariæ.* I have discussed these curious fossils at length in "Acadian Geology" and in the "Journal of the Geological Society of London," 1860. The following summary is from the last-mentioned paper:

a. As Prof. Williamson and the writer have shown, many of the *Sternbergia* piths belong to coniferous trees of the genus *Dadoxylon.*

b. A few specimens present multiporous tissue, of the type of *Dictyoxylon*, a plant of unknown affinities, and which, according to Williamson, has a *Sternbergia* pith.

c. Other examples show a true scalariform tissue, comparable with that of *Lepidodendron* or *Sigillaria*, but of finer texture. Corda has shown that plants of the type of the former genus (his *Lomatophloios*) had *Sternbergia* piths. Some plants of this group are by external characters loosely reckoned by botanists as ribless *Sigillariæ* (*Clathraria*); but I believe that they are not related even ordinally to that genus.

d. Many Carboniferous *Sternbergiæ* show structures identical with those described above as occurring in *Cordaites*, and also in some of the trees ordinarily reckoned as *Sigillariæ.*

Genus CARDIOCARPUM.

I have found at least eight species of these fruits in the Erian and Carboniferous of New Brunswick and Nova Scotia, all of which are evidently fruits of gymnospermous trees. They agree in having a dense coaly nucleus of appreciable thickness, even in the flattened specimens, and surrounded by a thin and veinless wing or margin. They have thus precisely the appearance of samaras of many existing forest-trees, some of which they also resemble in the outline of the margin, except that the wings of samaras are usually veiny. The character of the nucleus, and the occasional appearance in it of marks possibly representing cotyledons or embryos, forbids the supposition that they are spore-cases. They must have been fruits of phænogams. Whether they were winged fruits or seeds, or fruits with a pulpy envelope like those of cycads and some conifers, may be considered less certain. The not infrequent distortion of the margin is an argument in favour of the latter view, though this may also be supposed to have occurred in samaras partially decayed. On the other hand, their being always apparently flattened in one plane, and the nucleus being seldom, if ever, found denuded of its margin, are arguments in favour of their having been winged nutlets or seeds. Until recently I had regarded the latter view as more probable, and so stated the matter in the second edition of "Acadian Geology." I have, however, lately arrived at the conclusion that the *Cardiocarpa* of the type of *C. cornutum* were gymnospermous seeds, having two cotyledons embedded in an albumen and covered with a strong membranous or woody tegmen surrounded by a fleshy outer coat, and that the notch at the apex represents the foramen or micropyle of the ovule. The structure was indeed very similar to that of the seeds of *Taxus* and of *Salisburia.* With respect to some of the other species, however, especially those with very broad margins, it still appears likely that they were winged.

The *Cardiocarpa* were borne in racemes or groups, and it seems certain that some of them at least are the seeds of *Cordaites.* The association of some of them and of those of the next genus with *Sigillariæ* is so constant that I cannot doubt that some of them belong to plants of that genus, or possibly to taxine conifers. The great number of distinct species of these seeds, as compared with that of known trees which could have produced them, is very remarkable.

Genus TRIGONOCARPUM.

These are large angled nuts contained in a thick envelope, and showing internal structures resembling those of the seeds of modern

Taxineæ. There are numerous species, as well as allied seeds referred to the provisional genera *Rhabdocarpus* and *Carpolithes.* In *Trigonocarpum Hookeri* I have described the internal structure of one of those seeds, and many fine examples from the coal-field of St. Etienne, in France, have been described by Brongniart, so that their internal structure is very well known.

Genus ANTHOLITHES.

This is also a provisional genus, to include spikes of floral organs, some of which are known to have belonged to *Cordaites,* others probably to *Sigillariæ.*

Of Uncertain Affinities.
Family SIGILLARIACEÆ.

Under this name palæobotanists have included a great numbei of trees of the Carboniferous system, all of which are characterised by broad leaf-sears, with three vascular scars, and usually arranged in vertical rows, and by elongated three-nerved leaves, and roots of the stigmaria type—that is, with rounded pits, marking the attachment of rootlets spirally arranged. These trees, however, collected in the genus *Sigillaria* by arbitrary characters, which pass into those of the Lepidodendroid trees, have been involved in almost inextricable confusion, to disentangle which it will be necessary to consider : 1. The external characters of *Sigillariæ,* and trees confounded with them. 2. Subdivision of *Sigillariæ* by external markings. 3. The microscopic character of their stems. 4. What is known of their foliage and fruit.

1. *Characters of Sigillaroid and Lepidodendroid Trunks.*

It may be premised that the modes of determination in fossil botany are necessarily different from those employed in recent botany. The palæobotanist must have recourse to characters derived from the leaves, the scars left by their fall, and the internal structures of the stem. These parts, held in little esteem by botanists in describing modern plants, and much neglected by them, must hold the first place in the regard of the fossil botanist, whereas the fructification, seldom preserved, and generally obscure, is of comparatively little service. It is to be remarked also that in such generalised plants as those of the Palæozoic, remarkable rather for the development of the vegetative than of the reproductive organs, the former rise in importance as compared with their value in the study of modern plants.

In *Sigillariæ, Lepidodendra,* &c., the following surfaces of the stem may be presented to our inspection :

1. The outer surface of the epidermis without its leaves, but with the leaf-bases and leaf-scars more or less perfectly preserved. On this surface we may recognise: (1) Cellular swellings or projections of the bark to which the leaves are attached. These may be called *leaf-bases,* and they are sometimes very prominent. (2) The actual mark of the attachment of the leaf situated in the most prominent part of the leaf-base. This is the *leaf-scar.* (3) In the leaf-scar when well preserved we can see one or more minute punctures or prominences which are the points where the vascular bundles passing to the leaf found exit. These are the *vascular scars.*

When the leaves are attached, the leaf-scars and vascular scars cannot be seen, but the leaf-bases can be made out. Hence it is important, if possible, to secure specimens with and without the leaves. In flattened specimens the leaf-bases are often distorted by pressure and marked with furrows which must not be mistaken for true structural characters. The leaf-bases, which are in relief on the outer surface of the stem, of course appear as depressions on the mould in the containing rock, in which the markings often appear much more distinctly than on the plant itself.

2. The outer surface of the epidermis may have been removed or may be destroyed by the coarseness of the containing rock. In this case the leaf-bases are usually preserved on the surface of the outer or corky bark, but the leaf-scars and vascular scars have disappeared. This gives that condition of Lepidodendroid trees to which the name *Knorria* has been applied. When plants are in this state careful inspection may sometimes discover traces of the leaf-scars on portions of the stem, and thus enable the *Knorria* to be connected with the species to which it belongs.

3. The outer or corky bark may be removed, exposing the surface of the inner or fibrous and cellular bark, which in the plants in question is usually of great thickness. In this case neither the leaf-bases nor the scars are seen, but punctures or little furrows or ridges appear where the vascular bundles entered the inner bark. Specimens in this state are usually said to be decorticated, though only the outer bark is removed. It is often difficult to determine plants in this condition, unless some portion of the stem can be found still retaining the bark; but when care is taken in collecting, it will not infrequently be found that the true outer surface can be recovered from the containing rock, especially if a coaly layer representing the outer bark intervenes between this and the inner impression. Speci-

mens of this kind, taken alone, have been referred to the genera *Knorria, Bothrodendron,* and *Halonia.*

4. In some cases, though not frequently, the outer surface of the ligneous cylinder is preserved. It almost invariably presents a regularly striated or irregularly wrinkled appearance, depending upon the vertical woody wedges, or the positions of the medullary rays or vascular bundles. Specimens of this kind constituted some of the *Endogenites* of the older botanists, and the genus *Schizodendron* of Eichwald appears to include some of them. Many of them have also been incorrectly referred to Calamites.

5. In some cases the cast of the medullary cylinder or pith may alone be preserved. This may be nearly smooth or slightly marked by vertical striæ, but more usually presents a transverse striation, and not infrequently the transverse constrictions and septa characteristic of the genus Sternbergia. Loose *Sternbergiæ* afford little means of connecting them with the species to which they belong, except by the microscopic examination of the shreds of the ligneous cylinder which often cling to them.*

These facts being premised, the following general statements may be made respecting some of the more common Palæozoic genera, referring, however, principally to the perfect markings as seen on the epidermis:

Sigillaria.—Leaf-bases hexagonal or elongated, or confluent on a vertical ridge. Leaf-scars hexagonal or shield-shaped. Vascular scars three, the two lateral larger than the central. This last character is constant, depending on the fact that the leaves of Sigillaria have two or more vascular bundles. All so-called *Sigillariæ* having the central vascular scar largest, or only one vascular bundle, should be rejected from this genus. In young branches of branching *Sigillariæ* the leaf-scars sometimes appear to be spiral, but in the older stems they form vertical rows; interrupted, however, by transverse rows or bands of *fruit-scars,* each with a single large central vascular scar, and which have borne the organs of fructification. *Arthrocaulis* of McCoy is founded on this peculiarity.

Syringodendron.—Differs from Sigillaria in the leaf-scars, which are circular and with a single vascular bundle. It is a matter of doubt whether these plants were of higher rank than Sigillaria tending toward the pines, or of lower rank tending toward Cyclostigma. Their leaf-bases form vertical ridges.

Lepidodendron.—Leaf-bases rhombic, oval, or lanceolate, moder-

---

* See my paper, " Journal of Geological Society," vol. xxvii.

ately prominent. Leaf-scars rhombic or sometimes shield-shaped or heart-shaped, in the middle or upper part of the leaf-base. Vascular scars three—the middle one always largest and corresponding to the single nerve of the leaf; the lateral ones sometimes obsolete.

In older stems three modes of growth are observed. In some species the expansion of the bark obliterates the leaf-bases and causes the leaf-scars to appear separated by wide spaces of more or less wrinkled bark, which at length becomes longitudinally furrowed and simulates the ribbed character of Sigillaria. In others the leaf-bases grow in size as the trunk expands, so that even in large trunks they are contiguous though much larger than those on the branches. In others the outer bark, hardening at an early age, is incapable of either of the above changes, and merely becomes cleft into deep furrows in the old trunks.

Lepidophloios.—Leaf-bases transverse and prominent—often very much so. Leaf-scars transversely rhombic or oval with three vascular scars, the central largest. Leaves very long and one-nerved. Large strobiles or branchlets borne in two ranks or spirally on the sides of the stem, and leaving large, round scars (*cone-scars*), often with radiating impressions of the basal row of scales.

Species with long or drooping leaf-bases have been included in *Lepidophloios* and *Lomatophloios*. Species with short leaf-bases and cone-scars in two rows have been called *Ulodendron*, and some of them have been included in *Sigillaria* (sub-genus *Clathraria*). Decorticated stems are *Bothrodendron* and *Halonia*. Some of the species approach near to the last genus, especially to the Lepidodendra with rhombic leaf-bases like *L. tetragonum*.

Cyclostigma. — Leaf-bases undeveloped. Leaf-scars circular or horseshoe-shaped, small, with a central vascular scar. In old trunks of Cyclostigma the leaf-scars become widely separated, and sometimes appear in vertical rows. Young branches of Lepidodendron sometimes have the leaf-scars similar to those of Cyclostigma.

Leptophleum. — Leaf-bases flat, rhombic; leaf-scars obsolete; vascular scar single, central. The last two genera are characteristically Devonian.

In contradistinction from the trees above mentioned, the following general statements may be made respecting other groups:

In conifers the leaf-bases are usually elongated vertically, often scaly in appearance, and with the leaf-scar terminal and round, oval, or rhombic, and with a single well-marked vascular scar.

In Calamites, Calamodendron, and Asterophyllites the scars of the branchlets or leaves are circular or oval, with only a single vascular scar, and situated in verticils at the top of well-marked nodes of the stem.

In tree-ferns the leaf-bases are large and usually without a distinct articulating surface. The vascular bundles are numerous. Protopteris has rounded leaf-scars with a large horseshoe-shaped bundle of vessels above and small bundles below. Caulopteris has large elliptic or oval leaf-scars with vascular scars disposed concentrically. Palæopteris,* of Geinitz, has the leaf-scars transversely oval and the vascular bundles confluent in a transverse band with an appendage or outlying bundle below. Stemmatopteris has leaf-scars similar to those of Caulopteris, but the vascular bundles united into a horseshoe-shaped band.

2. Subdivision of Sigillariæ in Accordance with their Markings.

The following groups may be defined in this way; but, being based on one character only, they are of course in all probability far from natural:

1. *Sigillaria*, Brongniart. Type, *Sigillaria reniformis*, Brongniart, or *S. Brounii*, Dawson.—Stem with broad ribs, usually much broader than the usually oval or elliptical tripunctate areoles, but disappearing at base, owing to expansion of the stem. Leaves narrow, long, three-nerved.

2. *Rhytidolepis*, Sternberg. Type, *S. scutellata*, Brongniart.—Ribs narrow, and often transversely striate. Areoles large, hexagonal or shield-shaped, tripunctate. Leaves as in last group. Rings of rounded scars on the stems and branches mark attachment of fruit. It is possible that some of the smaller stems of this group may be branches of trees of group first.

3. *Syringodendron*, Sternberg. Type, *S. organum*, L. and H., *S. oculata*, Brongniart.—Stems ribbed; areoles small and round, and apparently with a single scar, or three closely approximated. These are rare, and liable to be confounded with decorticated examples of other groups; but I have some specimens which unquestionably represent the external surface.

4. *Favularia*, Sternberg. Type, *Sigillaria elegans* of Brongniart.—Leaf-bases hexagonal, or in young branches elliptical, in vertical rows, but without distinct ribs, except in old or decorticated stems. Fruit borne in verticils on the branches bearing transverse rows of rounded scars. Leaves somewhat broad and longitudinally striate.

---

* This name, preoccupied by Geinitz, has been inadvertently misapplied to the Devonian ferns of the genus *Archæopteris*.

5. *Leioderma*, Goldenberg.   Type, *S. Sydnensis*, Dawson. — Ribs obsolete.   Cortical and ligneous surfaces striate.   Vascular scars double, elongate longitudinally, and alike on cortical and inner surfaces.   Areoles in rows and distinct; stigmaria-roots striate, with small and distinct areoles.

6. *Clathraria*, Brongniart.   Type, *S. Menardi*, Brongniart.— Areoles hexagonal, not in distinct rows, but having a spiral appearance.   Some of the plants usually referred to this group are probably branches of *Favularia*.   Others are evidently fragments of plants of the genus *Lepidophloios*.

3. Internal Structures of Sigillaria-Stems.

I long ago pointed out, on the evidence of the external markings and mode of growth, that the stems of *Sigillariæ* must have been exogenous, and this conclusion has now been fully confirmed by the microscopic researches of Williamson, not only in the case of *Sigillariæ*, but of *Lepidodendra* and *Calamodendra* as well.   Confining myself to my own observations, three types of *Sigillariæ* are known to me by their internal structures, though I cannot certainly correlate all of these with the external markings referred to above.

1. *Diploxylon*, in which the stem consists of a small internal axis surrounded by a very thick inner bark and a dense outer cortex. A fine example from the South Joggins is thus described : [*]

" The axis of the stem is about six centimetres in its greatest diameter, and consists of a central pith-cylinder and two concentric coats of scalariform tissue.   The pith-cylinder is replaced by sandstone, and is about one centimetre in diameter.   The inner cylinder of scalariform tissue is perfectly continuous, not radiated, and about one millimetre in thickness.   Its vessels are somewhat crushed, but have been of large diameter.   Its outer surface, which readily separates from that of the outer cylinder, is striated longitudinally.   The outer cylinder, which constitutes by much the largest part of the whole, is also composed of scalariform tissue; but this is radially arranged, with the individual cells quadrangular in cross-section.   The cross-bars are similar on all the sides and usually simple and straight, but sometimes branching or slightly reticulated.   The wall intervening between the bars has extremely delicate longitudinal waving lines of ligneous lining, in the manner first described by Williamson as occurring in the scalariform tissue of certain *Lepidodendra*.   A few small radiating spaces, partially

---

[*] " Journal of the Geological Society of London," November, 1877.

occupied with pyrites, obscurely represent the medullary rays, which must have been very feebly developed. The radiating bundles passing to the leaves run nearly horizontally; but their structure is very imperfectly preserved. The stem being old and probably long deprived of its leaves, they may have been partially disorganised before it was fossilised. The outer surface of the axis is striated longitudinally, and in some places marked with impressions of tortuous fibres, apparently those of the inner bark. In the cross-section, where weathered, it shows concentric rings; but under the microscope these appear rather as bands of compressed tissue than as proper lines of growth. They are about twenty in number. This tree has an erect, ribbed trunk, twelve feet in height and fifteen inches in diameter, swelling to about two feet at the base.

2. *Favularia Type.*—This has been well described by Brongniart and by Renault,* and differs from the above chiefly in the fact that the outer exogenous woody zone is composed of reticulated instead of scalariform tissue, and the inner zone is of the peculiar form which I have characterised as pseudo-scalariform.

3. *Sigillaria Proper.*—This I have illustrated in my paper in the "Journal of the Geological Society" for May, 1871, and it appears to represent the highest and most perfect type of the larger ribbed *Sigillaria.* This structure I have described as follows, basing my description on a very fine axis found in an erect stem, and on the fragments of the woody axis found in the bases of other erect stems:

a. A dense cellular outer bark, usually in the state of compact coal—but when its structure is preserved, showing a tissue of thickened parenchymatous cells.

b. A very thick inner bark, which has usually in great part perished, or been converted into coal, but which, in old trunks, contained a large quantity of prosenchymatous tissue, very tough and of great durability. This "bast-tissue" is comparable with that of the inner bark of modern conifers, and constitutes much of the mineral charcoal of the coal-seams.

c. An outer ligneous cylinder, composed of wood-cells, either with a single row of large bordered pores,† in the manner of pines

---

* "Botanique Fossile," Paris, 1881.

† These are the same with the wood-cells elsewhere called discigerous tissue, and to which I have applied the terms uniporous and multiporous. The markings on the walls are caused by an unlined portion of the cell-wall placed in a disk or depression, and this often surrounded by an

and cycads, or with two, three, or four rows of such pores sometimes inscribed in hexagonal areoles in the manner of *Dadoxylon*. This woody cylinder is traversed by medullary rays, which are short, and composed of few rows of cells superimposed. It is also traversed by oblique radiating bundles of pseudo-scalariform tissue proceeding to the leaves. In some *Sigillariæ* this outer cylinder was itself in part composed of pseudo-scalariform tissue, as in Brongniart's specimen of *S. elegans;* and in others its place may have been taken by multiporous tissue, as in a case above referred to; but I have no reason to believe that either of these variations occurred in the typical ribbed species now in question. The woody fibres of the outer cylinder may be distinguished most readily from those of conifers, as already mentioned, by the thinness of their walls, and the more irregular distribution of the pores. Additional characters are furnished by the medullary rays and the radiating bundles of scalariform tissue when these can be observed.

d. An inner cylinder of pseudo-scalariform tissue. I have adopted the term pseudo-scalariform for this tissue, from the conviction that it is not homologous with the scalariform ducts of ferns and other acrogens, but that it is merely a modification of the discigerous wood-cells, with pores elongated transversely, and sometimes separated by thickened bars, corresponding to the hexagonal areolation of the ordinary wood-cells. A similar tissue exists in cycads, and is a substitute for the spiral vessels existing in ordinary exogens.

e. A large medulla, or pith, consisting of a hollow cylinder of cellular tissue, from which proceed numerous thin diaphragms towards the centre of the stem.

These structures of the highest type of *Sigillaria* are on the one hand scarcely advanced beyond those of *Calamopitus*, as described by Williamson, and on the other approach to those of *Cordaites*, as seen in specimens presented to me by Renault.

Finally, as to the fruit of *Sigillariæ*, I have no new facts to offer. The strobiles or spikes associated with these trees have been variously described as gymnospermous (Renault) or cryptogamous (Goldenberg and Williamson). I have never seen them in place. Two considerations, however, have always weighed with me in reference to this subject. One is the constant abundance of Trigonocarpa

---

hexagonal rim of thickened wall; but in all cases these structures are less pronounced than in *Dadoxylon*, and less regular in the walls of the same cell, as well as in different layers of the tissues of the axis.

and Cardiocarpa in the soil of the Sigillaria forests, as I have studied this at the South Joggins. The other is that the rings of fruit-scars on the branches of Sigillaria are homologous with leaf-scars, not with branches, and therefore should have borne single carpels and not cones or spikes of inflorescence. These are merely suggestions, but I have no doubt they will be vindicated by future discoveries, which will, I have no doubt, show that in the family *Sigillariaceæ* we have really two families, one possibly of gymnospermous rank, or at least approaching to this, the other allied to the Lepidodendra.

CRYPTOGAMIA.

(*Acrogenes.*)

Family LEPIDODENDREÆ; *Genus* LEPIDODENDRON, Sternberg.

These are arboreal Lycopods having linear one-nerved leaves, stems branching dichotomously, and with ovate or rhombic leaf-bases bearing rhombic leaf-scars, often very prominent. The fruit is in scaly strobiles, terminal or lateral, and there are usually, if not always, macrospores and microspores in each strobile. The young branches and stems have a central pith, a cylinder of scalariform tubes sending out ascending bundles to the leaves through a thick cellular and fibrous inner bark, and externally a dense cortex confluent with or consisting of the leaf-bases. Older stems have a second or outer layer of scalariform fibres in wedges with medullary rays, and strengthening the stem by a true exogenous growth, much as in the Diploxylon type of Sigillaria. The development of this exogenous cylinder is different in amount and rate in different species.[*] This different development of the exogenous axis is accompanied with appropriate external appearances in the stems, and the changes which take place in their markings. These are of three kinds. In some species the areoles, at first close together, become, in the process of the expansion of the stem, separated by intervening spaces of bark in a perfectly regular manner; so that in old stems, while widely separated, they still retain their arrangement, while in young stems they are quite close to one another. This is the case in *L. corrugatum.* In other species the leaf-scars or bases increase in size in the old stems, still retaining their forms and their contiguity to each other. This is the case in *L. undulatum,* and generally in those *Lepidodendra* which have large leaf-bases. In these species the

---

[*] See " Memoirs of Dr. Williamson," in " Philosophical Transactions," for ample details.

continued vitality of the bark is shown by the occasional production of lateral strobiles on large branches, in the manner of the modern red pine of America. In other species the areoles neither increase in size nor become regularly separated by growth of the intervening bark; but in old stems the bark splits into deep furrows, between which may be seen portions of bark still retaining the areoles in their original dimensions and arrangement. This is the case with *L. Pictoense.* This cracking of the bark no doubt occurs in very old trunks of the first two types, but not at all to the same extent.

As a type of Lepidodendron, I may describe one of the oldest Carboniferous species characteristic of the Lower Carboniferous in America, and corresponding to *L. Veltheimianum* of Europe.

LEPIDODENDRON CORRUGATUM, Dawson.—(See Fig. 43, *supra*.) "Quarterly Journal of Geological Society," vol. xv.; "Acadian Geology," page 451.

Habit of Growth.—Somewhat slender, with long branches and long, slender leaves having a tendency to become horizontal or drooping.

Markings of Stem.—Leaf-bases disposed in quincunx or spirally, elongate, ovate, acute at both ends, but more acute and slightly oblique at the lower end; most prominent in the upper third, and with a slight vertical ridge. Leaf-scars small, rounded, and showing only a single punctiform vascular scar. The leaf-scar on the outer surface is in the upper third of the base; but the obliquity of the vascular bundle causes it to be nearly central on the inside of the epidermis. In young succulent shoots the leaf-scars are contiguous and round as in Cyclostigma, without distinct leaf-bases. In this state it closely resembles *L. Olivieri*, Eichwald.[*]

In the ordinary young branches the leaf-scars are contiguous, and closely resemble those of *L. elegans*, Brongt. (Fig. 43 C). As the branches increase in diameter the leaf-scars slightly enlarge and sometimes assume a verticillate appearance (Fig. 43 D). As they still further enlarge they become separated by gradually increasing spaces of bark, marked with many waving striæ or wrinkles (Fig. 43 I, N). At the base of old stems the bark assumes a generally wrinkled appearance without distinct scars.

Knorria or Decorticated States.—Of these there is a great variety, depending on the state of preservation, and the particular longitudinal ridges. Fig. 43 D shows a form in which the vascular bundles appear as cylindrical truncate projections. Other forms show

---

[*] Lethæa Rossica, Plate Y, Figs. 12, 13.

the leaf-bases prominent, or have an appearance of longitudinal ribbing produced by the expansion of the bark.

Structure of Stem.—This is not perfectly preserved in any of
my specimens, but one flattened specimen shows a central medulla
with a narrow ring of scalariform vessels surrounding it, and constituting the woody axis. The structure is thus similar to that of *L.
Harcourtii*, which I regard as probably the same with the closely
allied European species *L. Veltheimianum.*

Leaves.—These are narrow, one-nerved, curving somewhat rapidly outward (Figs. 43, B, C, D). They vary from one to two inches
in length.

Roots.—I have not seen these actually attached, but they occur
very abundantly in the underclays of some erect forests of these
plants at Horton Bluff, and are of the character of *Stigmariæ* (Figs.
30, 31). In some of the underclays the long, flattened rootlets are ex-
cessively abundant, and show the mark of a central vascular bundle.

Fructification.—Cones terminal, short, with many small, acute
imbricate scales. Spore-cases globular, smooth (Fig. 43 C). On
the surface of some shales and sandstones at Horton there are innumerable round spore-cases of this tree about the size of mustard-seed
(Fig. 43 F). Large slabs are sometimes covered with these, and thin
layers of shale are filled with flattened specimens.

This is the characteristic species of the Lower Carboniferous coal-
measures, occurring in great profusion at Horton Bluff and its
vicinity, also at Sneid's Mills near Windsor, Noel and Five-Mile
River, at Norton Creek and elsewhere in New Brunswick (Matthew's
collection), and at Antigonish (Honeyman's collection).

I have received from the lowest Carboniferous beds of Ohio specimens of this species.[*] According to Rogers and Lesquereux similar
forms occur in the Vespertine of Pennsylvania and in the Lower
Carboniferous of Illinois. *L. Veltheimianum* of western Europe
and *L. glincanum* of Russia are closely allied Lower Carboniferous
species.[†]

A very different type is furnished by a new species from the
middle coal-formation of Clifton, New Brunswick.

LEPIDODENDRON CLIFTONENSE, Dawson. — *Habit of Growth.*—
Robust, with thick branches, and leaves several inches in length.
Terminal branches becoming slender, with shorter leaves.

---

[*] " Journal of Geological Society," November, 1862, p. 313.

[†] For comparisons of these see " Report on Plants of Lower Carboniferous of Canada," p. 21.

Markings of Stem.—Leaf-bases long oval, pointed at ends, enlarging with growth of stem. Leaf-scars central, rhombic, transverse.

Leaves.—One-nerved, acutely pointed, from four inches in length on the larger branches to one inch or less on the branchlets.

Fructification.—Cones large, cylindrical or long oval, with large scales of trigonal form, and not elongated but lying close to the surface. Borne on lateral, slender branchlets, with short leaves.

Genus LEPIDOPHLOIOS, Sternberg; ULODENDRON, L. and H.; LOMATOPHLOIOS, Corda.

Lepidophloios.—Under this generic name, established by Sternberg, I include those lycopodiaceous trees of the coal-measures which have thick branches, transversely elongated leaf-scars, each with three vascular points and placed on elevated or scale-like protuberances, long one-nerved leaves, and large lateral strobiles in vertical rows or spirally disposed. Their structure resembles that of *Lepidodendron*, consisting of a *Sternbergia* pith, a slender axis of large scalariform vessels, giving off from its surface bundles of smaller vessels to the leaves, a very thick cellular bark, and a thin dense outer bark, having some elongated cells or bast-tissue on its inner side. In these trees the exogenous outer cylinder is less developed than in the Lepidodendra, and is sometimes wanting in stems or branches of some thickness.

Regarding *L. laricinum* of Sternberg as the type of the genus, and taking in connection with this the species described by Goldenberg, and my own observations on numerous specimens found in Nova Scotia, I have no doubt that *Lomatophloios crassicaulis* of Corda, and other species of that genus described by Goldenberg, *Ulodendron* and *Bothrodendron* of Lindley, *Lepidodendron ornatissimum* of Brongniart, and *Halonia punctata* of Geinitz, all belong to this genus, and differ from each other only in conditions of growth and preservation. Several of the species of *Lepidostrobus* and *Lepidophyllum* also belong to *Lepidophloios*.

The species of *Lepidophloios* are readily distinguished from *Lepidodendron* by the form of the areoles, and by the round scars on the stem, which usually mark the insertion of the large strobiles, though in barren stems they may also have produced branches; still, the fact of my finding the strobiles *in situ* in one instance, the accurate resemblance which the scars bear to those left by the cones of the red pine when borne on thick branches, and the actual impressions of the radiating scales in some specimens, leave no doubt in my

mind that they are usually the marks of cones; and the great size of
the cones of *Lepidophloios* accords with this conclusion.

The species of *Lepidophloios* are numerous, and individuals are
quite abundant in the coal formation, especially toward its upper
part. Their flattened bark is frequent in the coal-beds and their
roofs, affording a thin layer of pure coal, which sometimes shows the
peculiar laminated or scaly character of the bark when other characters are almost entirely obliterated. The leaves also are nearly as
abundant as those of *Sigillaria* in the coal-shales. They can readily
be distinguished by their strong, angular mid-rib.

The markings of *Lepidophloios* may easily be mistaken for those
of the *Clathraria* type of *Sigillaria*. When the stem only is seen,
they can be distinguished by the length of the leaf-bases in *Lepidophloios*, and by the dominant central vascular scar; also by the
one-nerved and ribbed leaves. Where the large, round marks of the
cones are present, these are an infallible guide, never being present
in *Sigillaria*. As the cones grew on the upper sides of the branches,
the impression of the lower side often shows no cone-scars, or only
two lateral rows, whereas on the upper side of the same branch they
appear spirally arranged. I may describe as an example—

Lepidophloios Acadianus, Dawson. Leaf-bases broadly rhombic, or in old stems regularly rhombic, prominent, ascending, terminated by very broad rhombic scars having a central point and two
lateral obscure points. Outer bark laminated or scaly. Surface of
inner bark with single points or depressions. Leaves long, linear,
with a strong keel on one side, five inches or more in length. Cone-
scars sparsely scattered on thick branches, either in two rows or
spirally, both modes being sometimes seen on the same branch.
Scalariform axis scarcely an inch in diameter in a stem five inches
thick. Fruit, an ovate strobile with numerous acute scales covering
small globular spore-cases. This species is closely allied to *Ulodendron majus* and *Lepidophloios laricinus*, and presents numerous
varieties of marking. Coal-formation, Nova Scotia.

Family CALAMITEÆ; Genus CALAMITES, Suckow.

The plants of this genus are unquestionably allied to the modern *Equisetaceæ*, but excel these so much in variety of form and
structure, and are so capricious in their states of preservation, and so
liable to be mistaken for parts of plants generically different, that
they have given rise to much controversy. The following considerations will enable us to arrive at some certainty.

The genus *Calamites* was originally founded in the longitudinally ribbed and jointed stems so frequent in the coal-formation, and of which the common *C. Suckovii* is a typical form. The most perfect of these stems represent the outer surface immediately within the epidermis, in which case transverse lines or constrictions mark the nodes, and at the nodes there are rounded spots, sometimes indicating radial processes of the pith, first described by Williamson; in other cases, the attachment of branchlets, or in some specimens both. But some specimens show the outer surface of the epidermis, in which case the transverse nodal lines are usually invisible, though the scars of branchlets may appear. In still other examples the whole of the outer tissues have perished, and the so-called Calamite is a cast of the interior of the stem, showing merely longitudinal ribbing and transverse nodal constrictions. In studying these plants *in situ* in the erect Calamite brakes of the coal-formation of Nova Scotia, one soon becomes familiar with these appearances, but they are evidently unknown to the majority of palæo-botanists, though described in detail more than twenty years ago.

When the outer surface is preserved it is sometimes seen to bear verticils of long needle-like leaves (*C. Cistii*), or of branchlets with secondary whorls of similar leaves (*C. Suckovii* and *C. undulatus*). No Calamite known to me bears broad one-nerved leaves like those of *Asterophyllites* and *Annularia*, though the larger stems of these plants have been described as Calamites, and the term *Calamocladus* has been used to include both groups. The base of the Calamite stem usually terminates in a blunt point, and may be attached to a rhizome, or several stems may bud out from each other in a group or stool. The roots are long and cylindrical, sometimes branching. The fruit consists of spikes of spore-cases, borne in whorls and subtended by linear floral leaves. To these strobiles the name Calamostachys has been given.

Williamson has shown that the stem of Calamites consists of a central pith or cavity of large size surrounded by a cylinder consisting of alternate wedges of woody and cellular matter, with vertical canals at the inner sides of the wedges, and slender medullary rays. The thick cellular wedges intervening between the woody wedges he calls primary medullary rays; the smaller medullary rays in the wedges, secondary medullary rays. There is thus a highly complex exogenous stem based on the same principle with the stem of a common *Equisetum*, but with much greater strength and complexity.

Williamson has also shown that there are different sub-types of these stems. More especially he refers to the three following:

(a) *Calamites* proper, which has the woody wedges of scalariform or barred tissue with thin medullary rays, and the thick primary medullary rays are cellular.

(b) *Calamopitus* has reticulated or multiporous tissue in the woody wedges with medullary rays, and the primary medullary wedges are composed of elongated cells.

(c) *Calamodendron* has the woody wedges of barred tissue as in a, with medullary rays, but has the intervening medullary wedges of an elongated tissue approaching to woody fibre, and also with medullary rays.

To these I would add a fourth type, which I have described, from the coal-formation of Nova Scotia.*

(d) *Eucalamodendron* differs from *Calamodendron* in having true bordered pores or pseudo-scalariform slit-pored tissue, and corresponds to the highest type of calamitean stem.

I would also add that under a and b there are some species in which the woody cylinder is very thin in comparison to the size of the stem. In c and d the woody cylinder is thick and massive, and the stems are often large and nodose.

As an example of an ordinary Calamite in which the external surface and foliage are preserved, I may quote the following from my report on the "Flora of the Lower Carboniferous and Millstone Grit," 1873:

CALAMITES UNDULATUS, Brongniart.—This species is stated by Brongniart to be distinguished from the *C. Suckovii*, the characteristic Calamite of the middle coal-formation, by its undulated ribs marked with peculiar cellular reticulation. He suggests that it may be merely a variety of *C. Suckovii*, an opinion in which Schimper coincides; but since I have received large additional collections from Mr. Elder, containing not only the stems and branches, but also the leaves and rhizomes, I am constrained to regard it as a distinct though closely allied species.

The rhizomata are slender, being from one to two inches in diameter, and perfectly flattened. They are beautifully covered with a cellular reticulation on the thin bark, and show occasional round areoles marking the points of exit of the rootlets. I have long been familiar with irregular flattened stems thus reticulate, but have only recently been able to connect them with this species of Calamite.

The main stems present a very thin carbonaceous bark reticulated like the rhizomes. They have flat, broad ribs separated by deep

* "Quarterly Journal of the Geological Society," 1871.

and narrow furrows, and undulated in a remarkable manner even when the stems are flattened. This undulation is, however, perhaps an indication of vertical pressure while the plant was living, as it seems to have had an unusually thin and feeble cortical layer, and the undulations are apparently best developed in the lower part of the stem. At the nodes the ribs are often narrowed and gathered together, especially in the vicinity of the rounded radiating marks which appear to indicate the points of insertion of the branches. At the top of each rib we have the usual rounded areole, probably marking the insertion of a primary branchlet.

The branches have slender ribs and distant nodes, from which spring secondary branchlets in whorls, these bearing in turn small whorls of acicular leaflets much curved upward, and which are apparently round in cross section and delicately striate. They are much shorter than the leaves of *Calamites Suckovii*, and are less dense and less curved than those of *C. nodosus*, which I believe to be the two most closely allied species.

Lesquereux notices this species as characteristic of the lower part of the Carboniferous in Arkansas.

It will be observed that I regard the striated and ribbed stems not as internal axes, but as representing the outer surface of the plants. This was certainly the case with the present species and with *C. Suckovii* and *C. nodosus*. Other species, and especially those which belonged to Calamodendron, no doubt had a smooth or irregularly wrinkled external bark; but this gives no good ground for the manner in which some writers on this subject confound Calamites with Calamodendra, and both with Asterophyllites and Sphenophyllum. With this no one who has studied these plants, rooted in their native soils, and with their appendages still attached, can for a moment sympathise. One of the earliest geological studies of the writer was a bed of these erect Calamites, which he showed to Sir C. Lyell in 1844, and described in the "Proceedings of the Geological Society" in 1851, illustrating the habit of growth as actually seen well exposed in a sandstone cliff. Abundant opportunities of verifying the conclusions formed at that time have since occurred, the results of which have been summed up in the figures in Acadian Geology, which, though they have been treated by some botanists as merely restorations, are in reality representations of facts actually observed.

On these subjects, without entering into details, and referring for these to the elaborate discussions of Schimper, Williamson, and McNab, and to my paper on the subject, "Journal of the Geological Society," vol. xxvii, p. 54, I may remark:

1. That the aërial stems of ordinary Calamites had a thin cortical layer, with lacunæ and fibrous bundles and multiporous vessels—the whole not differing much from the structure of modern Equiseta.

2. Certain arborescent forms, perhaps allied to the true Calamites, as well as possibly the old underground stems of ordinary species,[*] assumed a thick-walled character in which the tissues resembled the wedges of an exogen, and abundance of pseudo-scalariform fibres were developed, while the ribbing of the external surface became obsolete or was replaced by a mere irregular wrinkling.

3. Sufficient discrimination has not been exercised in separating casts of the internal cavities of Calamites and Calamodendron from those representing other surfaces and the proper external surface.

4. There is no excuse for attributing to Calamites the foliage of Annularia, Asterophyllites, and Sphenophyllum, since these leaves have not been found attached to true Calamite stems, and since the structure of the stems of Asterophyllites as described by Williamson, and that of Sphenophyllum as described by the writer,[†] are essentially different from those of Calamites.

5. As the species above described indicates, good external characters can be found for establishing species of this genus, and these species are of value as marks of geological age.

Genus ARCHÆOCALAMITES, Sternberg.

This genus has been established to include certain Calamites of the Devonian and Lower Carboniferous, in which the furrows on the stem do not alternate at the nodes or joints, and the leaves in one species at least bifurcate. *C. radiatus*, Brongniart, is the typical species. In North America it occurs in the Erian, probably as low as the Middle Erian. In Europe it has so far been recognised in the Lower Carboniferous only. I have, however, seen stems from alleged Devonian beds in Devonshire which may have belonged to this species.

Family ASTEROPHYLLITEÆ; Genus ASTEROPHYLLITES, Brongniart.

Stems ribbed and jointed like the *Calamites*, but with inflated nodes and a stout internal woody cylinder, which has been described by Williamson. From the joints proceeded whorls of leaves or of branchlets, bearing leaves which differed from those of *Calamites* in their having a distinct middle rib or vein. The fructification con-

---

[*] Williamson, "Transactions of the Royal Society." McNab, in "Proceedings of the Edinburgh Botanical Society."

[†] "Journal of the Geological Society," 1866.

sisted of long slender cones or spikes, having whorls of scales bearing the spore-cases. Some authors speak of *Asterophyllites* as only branches and leaves of *Calamites;* but though at first sight the resemblance is great, a close inspection shows that the leaves of *Asterophyllites* have a true midrib, which is wanting in *Calamites.*

Genus ANNULARIA.—It is perhaps questionable whether these plants should be separated from *Asterophyllites.* The distinction is that they produce branches in pairs, and that their whorls of leaves are one-sided and usually broader than those of *Asterophyllites,* and united into a ring at their insertion on the stem. One little species, *A. sphenophylloides,* is very widely distributed.

PINNULARIA—a provisional genus—includes slender roots or stems branching in a pinnate manner, and somewhat irregularly. They are very abundant in the coal shales, and were probably not independent plants, but aquatic roots belonging to some of the plants last mentioned. The probability of this is farther increased by their resemblance in miniature to the roots of *Calamites.* They are always flattened, but seem originally to have been round, with a slender thread-like axis of scalariform vessels, enclosed in a soft, smooth, cellular bark.

Family RHIZOCARPEÆ; *Genus* SPHENOPHYLLUM.

Leaves in whorls, wedge-shaped, with forking veins. Fructification on spikes, with verticils of sporocarps. These plants are by some regarded as allied to the *Calamiteæ* and *Asterophylliteæ,* by others as a high grade of Rhizocarps of the type of Marsilia. The stem had a star-shaped central bundle of scalariform or reticulato-scalariform vessels.

Genus SPORANGITES. (*Sporocarpon,* Williamson.)

Under this name we may provisionally include those rounded spherical bodies found in the coal and its accompanying beds, and also in the Erian, which may be regarded as Macrospores or Sporocarps of Protosalvinia, or other Rhizocarpean plants akin to those described above in Chapter III, which see for description.

Genus PROTOSALVINIA.—Under this we include sporocarps allied to those of Salvinia, as described in Chapter III.

Family FILICES.

Under this head I shall merely refer to a few groups of special interest, and to the provisional arrangement adopted for the fronds of ferns when destitute of fructification.

The external appearances of trunks of tree-ferns have been already referred to.

With respect to tree ferns, the oldest known examples are those from the Middle Devonian of New York and Ohio, which I have described in the "Journal of the Geological Society," 1871 and 1881. As these are of some interest, I have reproduced their descriptions in a note appended to Chapter III, which see.

The other forms most frequently occurring in the Carboniferous are *Caulopteris*, *Palæopteris*, and *Megaphyton*.* Stems showing merely masses of aërial roots are known by the name *Psaronius*.

With reference to the classification of Palæozoic ferns, this has hitherto been quite arbitrary, being based on mere form and venation of fronds, but much advance has recently been made in the knowledge of their fructification, warranting a more definite attempt at classification. The following are provisional genera usually adopted:

1. *Cyclopteris*, Brongniart.—Leaflets more or less rounded or wedge-shaped, without midrib, the nerves spreading from the point of attachment. This group includes a great variety of fronds evidently of different genera, were their fructification known; and some of them probably portions of fronds, the other parts of which may be in the next genus.

2. *Neuropteris*, Brongniart.—Fronds pinnate, and with the leaflets narrowed at the base; midrib often not distinct, and disappearing toward the apex. Nervures equal, and rising at an acute angle. Ferns of this type are among the most abundant in the coal-formation.

3. *Odontopteris*, Brongniart.—In these the frond is pinnate, and the leaflets are attached by their whole base, with the nerves either proceeding wholly from the base, or in part from an indistinct midrib, which soon divides into nervures.

4. *Dictyopteris*, Gutbier.—This is a beautiful style of fern, with leaflets resembling those of *Neuropteris*, but the veins arranged in a network of oval spaces. Only a few species are known in the coal-formation.

5. *Lonchopteris*, Brongniart.—Ferns with netted veins like the above, but with a distinct midrib, and the leaflets attached by the whole base. Of this, also, we can boast but few species.

6. *Sphenopteris*, Brongniart.—These are elegant ferns, very numerous in species, and most difficult to discriminate. Their most

---

* See my "Acadian Geology," also below.

distinctive characters are leaflets narrowed at the base, often lobed, and with nervures dividing in a pinnate manner from the base.

7. *Phyllopteris*, Brongniart.—These are pinnate, with long lanceolate pinnules, having a strong and well-defined midrib, and nerves proceeding from it very obliquely, and dividing as they proceed toward the margin. The ferns of this genus are for the most part found in formations more recent than the Carboniferous; but I have referred to it, with some doubt, one of our species.

8. *Alethopteris*, Brongniart.—This genus includes many of the most common coal-formation ferns, especially the ubiquitous *A. lonchitica*, which seems to have been the common brake of the coal-formation, corresponding to *Pteris aquilina* in modern Europe and America. These are brake-like ferns, pinnate, with leaflets often long and narrow, decurrent on the petiole, adherent by their whole base, and united at base to each other. The midrib is continuous to the point, and the nervures run off from it nearly at right angles. In some of these ferns the fructification is known to have been marginal, as in *Pteris*.

9. *Pecopteris*, Brongniart.—This genus is intermediate between the last and *Neuropteris*. The leaflets are attached by the whole base, but not usually attached to each other; the midrib, though slender, attains to the summit; the nervures are given off less obliquely than in *Neuropteris*. This genus includes a large number of our most common fossil ferns.

10. *Beinertia*, Goeppert.—A genus established by Goeppert for a curious Pecopteris-like fern, with flexuous branching oblique nervures becoming parallel to the edge of the frond.

11. *Hymenophyllites*, Goeppert.—These are ferns similar to *Sphenopteris*, but divided at the margin into *one-nerved* lobes, in the manner of the modern genus *Hymenophyllum*.

12. *Palæopteris*, Geinitz.—This is a genus formed to include certain trunks of tree-ferns with oval transverse scars of leaves.

13. *Caulopteris*, Lindley and Hutton.—Is another genus of fossil trunks of tree-ferns, but with elongate scars of leaves.

14. *Psaronius*, Cotta.—Includes other trunks of tree-ferns with alternate scars or thick scales, and ordinarily with many aërial roots grouped round them, as in some modern tree-ferns.

15. *Megaphyton*, Artis.—Includes trunks of tree-ferns which bore their fronds, which were of great size, in two rows, one on each side of the stem. These were very peculiar trees, less like modern ferns than any of the others. My reasons for regarding them as ferns are stated in the following extract from a recent paper:

"Their thick stems, marked with linear scars and having two rows of large depressed areoles on the sides, suggest no affinities to any known plants. They are usually ranked with *Lepidodendron* and *Ulodendron*, but sometimes, and probably with greater reason, are regarded as allied to tree-ferns. At the Joggins a very fine species (*M. magnificum*) has been found, and at Sydney a smaller species (*M. humile*); but both are rare and not well preserved. If the large scars bore cones and the smaller bore leaves, then, as Brongniart remarks, the plant would much resemble *Lepidophloios*, in which the cone-scars are thus sometimes distichous. But the scars are not round and marked with radiating scales as in *Lepidophloios;* they are reniform or oval, and resemble those of tree-ferns, for which reason they may be regarded as more probably leaf-scars; and in that case the smaller linear scars would indicate ramenta, or small aërial roots. Further, the plant described by Corda as *Zippea disticha* is evidently a *Megaphyton*, and the structure of that species is plainly that of a tree-fern of somewhat peculiar type. On these grounds I incline to the opinion of Geinitz that these curious trees were allied to ferns, and bore two rows of large fronds, the trunks being covered with coarse hairs or small aërial roots. At one time I was disposed to suspect that they may have crept along the ground; but a specimen from Sydney shows the leaf-stalks proceeding from the stem at an angle so acute that the stem must, I think, have been erect. From the appearance of the scars it is probable that only a pair of fronds were borne at one time at the top of the stem; and, if these were broad and spreading, it would be a very graceful plant. To what extent plants of this type contributed to the accumulation of coal I have no means of ascertaining, their tissues in the state of coal not being distinguishable from those of ferns and *Lycopodiaceæ*."

16. For descriptions of the genus *Archæopteris* and other Erian ferns, see Chapter III.

CHAPTER V.

THE FLORA OF THE EARLY MESOZOIC.

GREAT physical changes occurred at the close of the Carboniferous age. The thick beds of sediment that had been accumulating in long lines along the primitive continents had weighed down the earth's crust. Slow subsidence had been proceeding from this cause in the coal-formation period, and at its close vast wrinklings occurred, only surpassed by those of the old Laurentian time. Hence in the Appalachian region of America we have the Carboniferous beds thrown into abrupt folds, their shales converted into hard slates, their sandstones into quartzite and their coals into anthracite, and all this before the deposition of the Triassic Red Sandstones which constitute the earliest deposit of the great succeeding Mesozoic period. In like manner the coal-fields of Wales and elsewhere in western Europe have suffered similar treatment, and apparently at the same time.

This folding is, however, on both sides of the Atlantic limited to a band on the margin of the continents, and to certain interior lines of pressure, while in the middle, as in Ohio and Illinois in America, and in the great interior plains of Europe, the coal-beds are undisturbed and unaltered. In connection with this we have an entire change in the physical character of the deposits, a great elevation of the borders of the continents, and probably a considerable deepening of the seas, leading to the establishment of general geographical conditions which still remain, though they have been temporarily modified by subsequent subsidences and re-elevations.

Along with this a great change was in progress in vegetable and animal life. The flora and fauna of the Palæozoic gradually die out in the Permian and are replaced in the succeeding Trias by those of the Mesozoic time. Throughout the Permian, however, the remains of the coal-formation flora continue to exist, and some forms, as the *Calamites*, even seem to gain in importance, as do also certain types of coniferous trees. The Triassic, as well as the Permian, was marked by physical disturbances, more especially by great volcanic eruptions discharging vast beds and dykes of lava and layers of volcanic ash and agglomerate. This was the case more especially along the margins of the Atlantic, and probably also on those of the Pacific. The volcanic sheets and dykes associated with the Red Sandstones of Nova Scotia, Connecticut, and New Jersey are evidences of this.

At the close of the Permian and beginning of the Trias, in the midst of this transition time of physical disturbance, appear the great reptilian forms characteristic of the age of reptiles, and the earliest precursors of the mammals, and at this time the old Carboniferous forms of plants finally pass away, to be replaced by a flora scarcely more advanced, though different, and consisting of pines, cycads, and ferns, with gigantic equiseta, which are the successors of the genus *Calamites*, a genus which still survives in the early Trias. Of these groups the conifers, the ferns, and the equiseta are already familiar to us, and, in so far as they are concerned, a botanist who had studied the flora of the Carboniferous would have found himself at home in the succeeding period. The cycads are a new introduction. The whole, however, come within the limits of the cryptogams and the gymnosperms, so that here we have no advance.*

---

* Fontaine's "Early Mesozoic Flora of Virginia" gives a very good summary of this flora in America.

As we ascend, however, in the Mesozoic, we find new and higher types. Even within the Jurassic epoch, the next in succession to the Trias, there are clear indications of the presence of the endogens, in species allied to

Fig. 64.—Jurassic vegetation. Cycads and pines. (After Saporta.)

the screw-pines and grasses; and the palms appear a little later, while a few exogenous trees have left their remains in the Lower Cretaceous, and in the Middle and Upper Cretaceous these higher plants come in abundantly and in generic forms still extant, so that the dawn of the modern flora belongs to the Middle and Upper

N

Cretaceous. It will thus be convenient to confine ourselves in this chapter to the flora of the earlier Mesozoic.

Passing over for the present the cryptogamous plants
already familiar in older deposits, we may notice the new
features of gymnospermous and phænogamous life, as they
present themselves in this earlier part of the great reptilian age, and as they extended themselves with remarkable uniformity in this period over all parts of the world.
For it is a remarkable fact that, if we place together in
our collections fossil plants of this period from Australia,
India, China, Siberia, Europe, or even from Greenland,
we find wonderfully little difference in their aspect. This
uniformity we have already seen prevailed in the Palæozoic flora ; and it is perhaps equally marked in that of
the Mesozoic. Still we must bear in mind that some
of the plants of these periods, as the ferns and pines,
for example, are still
world-wide in their
distribution ; but this
does not apply to oth-
ers, more especially
the cycads (Fig. 65).

The cycads consti-
tute a singular and ex-
ceptional type in the
modern world, and
are limited at present
to the warmer cli-
mates, though very
generally distributed
in these, as they oc-
cur in Africa, India,

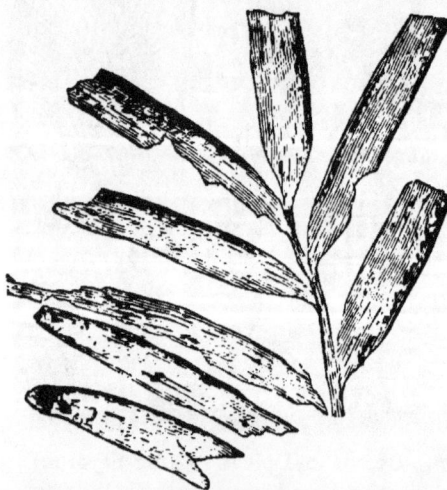

Fig. 65.—*Podozamites lanceolatus*, Sternb.
L. Cretaceous.

Japan, Australia, Mexico, Florida, and the West Indies.
In the Mesozoic age, however, they were world-wide in
their distribution, and are found as far north as Greenland, though most of the species found in the Cretaceous

of that country are of small size, and may have been of
low growth, so that they may have been protected by the
snows of winter.   The cycads have usually simple or un-
branching stems, pinnate leaves borne in a crown at top,
and fruits which, though somewhat various in structure
and arrangement, are all of the simpler form of gymnospermous type.   The stems are exogenous in structure,
but with slender wood and thick bark, and barred tissue,
or properly as tissue intermediate between this and the
disc-bearing fibres of the pines.

Though the cycads have a considerable range of organisation and of fructification, and though some points
in reference to the latter might assign them a higher
place, on the whole they seem to occupy a lower position
than the conifers or the cordaiteæ of the Carboniferous.
In the Carboniferous some of the fern-like leaves assigned
to the genus *Noeggerathia* have been shown by Stur and
Weiss to have been gymnosperms, probably allied to
cycads, of which they may be regarded at least as precursors.   Thus the cycadean type does not really constitute an advance in grade of organisation in the Mesozoic,
any further than that, in the period now in question, it
becomes much more developed in number and variety of
forms.   But the conifers would seem to have had preced-
ence of it for a long time in the Palæozoic, and it replaces
in the Mesozoic the *Cordaites,* which in many respects
excelled it in complexity.

The greater part of the cycads of the Mesozoic age
would seem to have had short stems and to have constituted the undergrowth of woods in which conifers attained to greater height.   An interesting case of this is
the celebrated dirt-bed of the quarries of the Isle of Port-
land, long ago described by Dean Buckland.   In this
fossil soil trunks of pines, which must have attained to
great height, are interspersed with the short, thick stems
of cycads, of the genus named *Cycadoidea* by Buckland,.

and which from their appearance are called "fossil birds' nests" by the quarrymen. Some, however, must have attained a considerable height so as to resemble palms.

The cycads, with their simple, thick trunks, usually marked with rhombic scars, and bearing broad spreading crowns of large, elegantly formed pinnate leaves, must have formed a prominent part of the vegetation of the northern hemisphere during the whole of the Mesozoic period. A botanist, had there been such a person at the time, would have found this to be the case everywhere from the equator to Spitzbergen, and probably in the southern hemisphere as well, and this throughout all the long periods from the Early Trias to the Middle Cretaceous. In a paper published in the "Linnæan Transactions" for 1868, Dr. Carruthers enumerates twenty species of British Mesozoic cycads, and the number might now be considerably increased.

The pines present some features of interest. We have already seen their connection with the broad-leaved *Cordaites*, and in the Permian there are some additional types of broad-leaved coniferæ. In the Mesozoic we have great numbers of beautiful trees, with those elegant fan-shaped leaves characteristic of but one living species, the *Salisburia*, or gingko-tree of China. It is curious that this tree, though now limited to eastern Asia, will grow, though it rarely fruits, in most parts of temperate Europe, and in America as far north as Montreal, and that in the Mesozoic period it occupied all these regions, and even Siberia and Greenland, and with many and diversified species (Fig. 66).

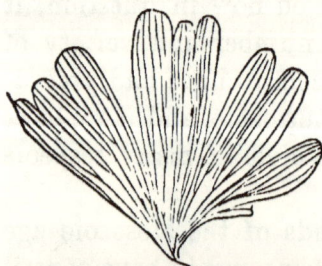

FIG. 66.—*Salisburia* (Gingko) *Sibirica*, Heer. L. Cretaceous, Siberia and North America.

Salisburia belongs to the yews, but an equally curious fact applies to the cypresses. The genus *Sequoia*, limited at present to two species, both Californian, and one of them the so-called "big tree," celebrated for the gigantic size to which it attains, is represented by species found as far back at least as the Lower Cretaceous, and in every part of the northern hemi-sphere.* It seems to have thriven in all these regions throughout the Mesozoic and early Kainozoic, and then to have disappeared, leaving only a small remnant to represent it in modern days. A number of species have been described from the Mesozoic and Tertiary, all of them closely related to those now existing (Fig. 67).

The following notice of these trees is for the most part translated, with some modifications and abridgment, from a paper read by the late Prof. Heer before the Botanical Section of the Swiss Natural History Society :

The name itself deserves consideration. It is that of an Indian of the Cherokee tribe, Sequo Yah, who invented an alphabet without any aid from the outside world of culture, and taught it to his tribe by writing it upon

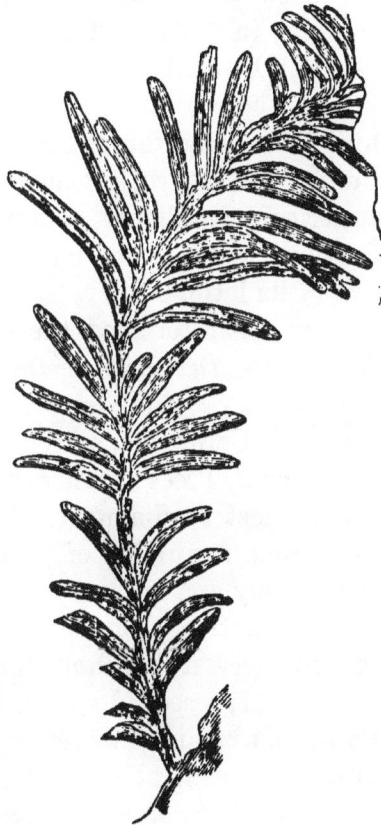

FIG. 67.—*Sequoia Smithiana*, Heer. L. Cretaceous.

---

* In the Eocene of Australia.

leaves. This came into general use among the Chero-
kees, before the white man had any knowledge of it ; and
afterward, in 1828, a periodical was published in this
character by the missionaries. Sequo Yah was banished
from his home in Alabama, with the rest of his tribe, and
settled in New Mexico, where he died in 1843.

When Endlicher was preparing his synopsis of the
conifers, in 1846, and had established a number of new
genera, Dr. Jacbon Tschudi, then living with Endlicher,
brought before his notice this remarkable man, and asked
him to dedicate this red-wooded tree to the memory of a
literary genius so conspicuous among the red men of
America. Endlicher consented to do so, and only en-
deavored to make the name pronounceable by changing
two of its letters.

Endlicher founded the genus on the redwood of the
Americans, *Taxodium sempervirens* of Lamb; and named
the species *Sequoia sempervirens.* These trees form large
forests in California, which extend along the coast as far
as Oregon. Trees are there met with of 300 feet in height
and 20 feet in diameter. The seeds have been brought
to Europe a number of years ago, and we already see in
upper Italy and around the Lake of Geneva, and in England, high trees ; but, on the other hand, they have not
proved successful around Zurich.

In 1852, a second species of Sequoia was discovered in
California, which, under the name of big tree, soon attained a considerable celebrity. Lindley described it, in
1853, as *Wellingtonia gigantea;* and, in the following
year, Decaisne and Torrey proved that it belonged to
Sequoia, and that it accordingly should be called *Sequoia
gigantea.*

While the *Sequoia sempervirens,* in spite of the de-
structiveness of the American lumbermen, still forms
large forests along the coast, the *Sequoia gigantea* is confined to the isolated clumps which are met with inland at

a height of 5,000 to 7,000 feet above sea-level, and are much sought after by tourists as one of the wonders of the country. Reports came to Europe concerning the largest of them which were quite fabulous, but we have received accurate accounts of them from Prof. Whitney. The tallest tree measured by him has a height of 325 feet, and in the case of one of the trees the number of the rings of growth indicated an age of about 1,300 years. It had a girth of 50 to 60 feet.

We know only two living species of *Sequoia*, both of which are confined to California. The one (*S. sempervirens*) is clothed with erect leaves, arranged in two rows, very much like our yew-tree, and bears small, round cones ; the other (*S. gigantea*) has smaller leaves, set closely against the branches, giving the tree more the appearance of the cypress. The cones are egg-shaped, and much larger. These two types are therefore sharply defined.

Both of these trees have an interesting history. If we go back into the Tertiary, this same genus meets us with a long array of species. Two of these species correspond to those living at present : the *S. Langsdorfii* to the *S. sempervirens*, and the *S. Couttsiæ* to the *S. gigantea.**
But, while the living species are confined to California, in the Tertiary they are spread over several quarters of the globe.

Let us first consider the *Sequoia Langsdorfii*. This was first discovered in the lignite of Wetterau, and was described as *Taxites langsdorfii*. Heer found it in the upper Rhone district, and there lay beside the twigs the remains of a cone, which showed that the *Taxites Langsdorfii* of Brongniart belonged to the Californian genus *Sequoia* established by Endlicher. He afterward

---

* *S. Couttsiæ* has leaves like *S. gigantea*, and cones like those of *S. sempervirens*.

found much better preserved cones, together with seeds, along with the plants of east Greenland, which fully confirmed the determination. At Atanekerdluk in Greenland (about 70° north latitude) this tree is very common. The leaves, and also the flowers and numerous cones, leave no doubt that it stands very near to the modern redwood. It differs from it, however, in having a much larger number of scales in the cone. The tree is also found in Spitzbergen at nearly 78° north latitude, where Nordenskiöld has collected, at Cape Lyell, wonderfully preserved branches. From this high latitude the species can be followed down through the whole of Europe as far as the middle of Italy (at Senegaglia, Gulf of Spezia). In Asia, also, we can follow it to the steppes of Kirghisen, to Possiet, and to the coast of the Sea of Japan, and across to Alaska and Sitka. It is recognized by Mr. Starkie Gardner as one of the species found in the Eocene of Mull in the Hebrides.* It is thus known in Europe, Asia, and America, from 43° to 78° north latitude, while its most nearly related living species, perhaps even descended from it, is now confined to California.

With this *S. Langsdorfii*, three other Tertiary species are nearly related (*S. brevifolia*, Hr., *S. disticha*, Hr., and *S. Nordenskiöldi*, Hr.). These have been met with in Greenland and Spitzbergen, and one of them has lately been found in the United States. Three other species, in addition to these, have been described by Lesquereux, which appear to belong to the group of the *S. Langsdorfii*, viz., *S. longifolia*, Lesq., *S. angustifolia*, and *S. acuminata*, Lesq. Several species also occur in the Cretaceous and Eocene of Canada.

These species thus answer to the living *Sequoia sempervirens ;* but we can also point to Tertiary represen-

---

* It is *Fareites Campbelli* of Forbes.

tatives of the *S. gigantea*. Their leaves are stiff and
sharp-pointed, are thinly set round the branches, and lie
forward in the same way : the egg-shaped cones are in
some cases similar.

There are, however, in the early Tertiary six species,
which fill up the gap between *S. sempervirens* and *S.
gigantea*. They are the *S. Couttsiæ*, *S. affinis*, Lesq.,
S. imbricata, Hr., *S. sibirica*, Hr., *S. Heerii*, Lesq., and
S. biformis, Lesq. Of these, *S. Couttsiæ*, Hr., is the
most common and most important species. It has short
leaves, lying along the branch, like *S. gigantea*, and
small, round cones, like *S. Langsdorfii* and *sempervirens*.
Bovey Tracey in Devonshire has afforded splendid specimens of cones, seeds, and twigs, which have been described
in the "Philosophical Transactions." More lately, Count
Saporta has described specimens of cones and twigs from
Armissan. Specimens of this species have also been found
in the older Tertiary of Greenland, so that it must have
had a wide range. It is very like to the American *S.
affinis*, Lesq.

In the Tertiary there have been already found fourteen
well-marked species, which thus include representatives
of the two living types, *S. sempervirens* and *S. gigantea*.

We can follow this genus still further back. If we go
back to the Cretaceous age, we find ten species, of which
five occur in the Urgon of the Lower Cretaceous, two in
the Middle, and three in the Upper Cretaceous. Among
these, the Lower Cretaceous exhibits the two types of the
Sequoia sempervirens and *S. gigantea*. To the former
the *S. Smithiana* answers, and to the latter, the *Reichenbachii*, Gein. The *S. Smithiana* stands indeed uncom-
monly near the *S. Langsdorfii*, both in the appearance of
the leaves on the twigs and in the shape of the cones.
These are, however, smaller, and the leaves do not become
narrower toward the base. The *S. pectina*, Hr., of the
Upper Cretaceous, has its leaves arranged in two rows, and

presents a similar appearance. The *S. Reichenbachii* is a type more distinct from those now living and those in the Tertiary. It has indeed stiff, pointed leaves, lying forward, but they are arcuate, and the cones are smaller. This tree has been known for a long time, and it serves in the Cretaceous as a guiding star, which we can follow from the Urgonian of the Lower Cretaceous up to the Cenomanian. It is known in France, Belgium, Bohemia, Saxony, Greenland, and Spitzbergen (also in Canada and the United States). It has been placed in another genus —Geinitzia—but we can recognise, by the help of the cones, that it belongs to Sequoia.

Below this, there is found in Greenland a nearly related species, the *S. ambigua*, Hr., of which the leaves are shorter and broader, and the cones round and somewhat smaller.

The connecting link between *S. Smithiana* and *Reichenbachii* is formed by *S. subulata*, Hr., and *S. rigida*, Hr., and three species (*S. gracilis*, Hr., *S. fastigiata* and *S. Gardneriana*, Carr.), with leaves lying closely along the branch, and which come very near to the Tertiary species *S. Couttsiæ*. We have therefore in the Cretaceous quite an array of species, which fill up the gap between the *S. sempervirens* and *gigantea*, and show us that the genus Sequoia had already attained a great development in the Cretaceous. This was still greater in the Tertiary, in which it also reached its maximum of geographical distribution. Into the present world the two extremes of the genus have alone continued; the numerous species forming its main body have fallen out in the Tertiary.

If we look still further back, we find in the Jura a great number of conifers, and, among them, we meet in the genus Pinus with a type which is highly developed, and which still survives; but for Sequoia we have till now looked in vain, so that for the present we can not place the rise of the genus lower than the Urgonian of the Cretaceous, however remarkable we may think it that in that period it should have developed into so many species ; and it is still more surprising that two species already make their appearance which approach so near to the living *Sequoia sempervirens* and *S. gigantea.*

Altogether, we have become acquainted, up to the present time, with twenty-six species of Sequoia. Fourteen of these species are found in the Arctic zone, and have been described and figured in the "Fossil Flora of the Arctic Regions." Sequoia has been recognised by Ettingshausen even in Australia, but there in the Eocene.

This is, perhaps, the most remarkable record in the whole history of vegetation. The Sequoias are the giants of the conifers, the grandest representatives of the family, and the fact that, after spreading over the whole northern hemisphere and attaining to more than twenty specific forms, their decaying remnant should now be confined to one limited region in western America and to two species constitutes a sad memento of departed greatness.* The small remnant of *S. gigantea* still, however, towers above all competitors, as eminently the "big trees" ; but, had they and the allied species failed to escape the Tertiary continental submergences and the disasters of the glacial period, this grand genus would have been to us an extinct type. In like manner the survival of the single gingko of eastern Asia alone enables us to understand that great series of taxine trees with fern-like leaves of which it is the sole representative.

Besides these peculiar and now rare forms, we have in the Mesozoic many others related closely to existing yews, cypresses, pines, and spruces, so that the conifers were probably in greater abundance and variety than they are at this day.

---

* The writer has shown that much of the material of the great lignite beds of the Canadian Northwest consists of wood of *Sequoia* of both the modern types.

In this period, also, we find the earliest representatives of the endogenous plants. It is true that some plants found in the coal-formation have been doubtfully referred to these, but the earliest certain examples would seem to be some bamboo-like and screw-pine-like plants occurring in the Jurassic rocks. Some of these are, it is true, doubtful forms, but of others there seems to be no question. The modern *Pandanus* or screw-pine of the tropical regions, which is not a pine, however, but a humble relation of the palms, is a stiffly branching tree, of a candelabra-like form, and with tufts of long leaves on its branches, and nuts or great hard berries for fruit, borne sometimes in large masses, and so protected as to admit of their drifting uninjured on the sea. The stems are supported by masses of aërial roots like those which strengthen the stems of tree-ferns. These structures and habits of growth fit the Pandanus for its especial habitat on the shores of tropical islands, to which its masses of nuts are drifted by the winds and currents, and on whose shores it can establish itself by the aid of its aërial roots.

Some plants referred to the cycads have proved veritable botanical puzzles. One of these, the *Williamsonia gigas* of the English oölite, originally discovered by my friend Dr. Williamson, and named by him *Zamia gigas,* a very tall and beautiful species, found in rocks of this age in various parts of Europe, has been claimed by Saporta for the Endogens, as a plant allied to *Pandanus.* Some other botanists have supposed the flowers and fruits to be parasites on other plants, like the modern *Rafflesia* of Sumatra, but it is possible that after all it may prove to have been an aberrant cycad.

The tree-palms are not found earlier than the Middle Cretaceous, where we shall notice them in the next chapter. In like manner, though a few Angiosperms occur in rocks believed to be Lower or Lower Middle Cretaceous in Greenland and the northwest territory of Canada, and in Virginia, these are merely precursors of those of the Upper Cretaceous, and are not sufficient to redeem the earlier Cretaceous from being a period of pines and cycads.

On the whole, this early Mesozoic flora, so far as known to us, has a monotonous and mean appearance. It no doubt formed vast forests of tall pines, perhaps resembling the giant Sequoias of California; but they must for the most part have been dark and dismal woods, probably tenanted by few forms of life, for the great reptiles of this age must have preferred the open and sunny coasts, and many of them dwelt in the waters. Still we must not be too sure of this. The berries and nuts of the numerous yews and cycads were capable of affording much food. We know that in this age there were many great herbivorous reptiles, like *Iguanodon* and *Hadrosaurus,* some of them fitted by their structure to feed upon the leaves and fruits of trees. There were also several kinds of small herbivorous mammals, and much insect life, and it is likely that few of the inhabitants of the Mesozoic woods have been preserved as fossils. We may yet have much to learn of the inhabitants of these forests of ferns, cycads, and pines. We must not forget in this connection that in the present day there are large islands, like New Zealand, destitute of mammalia, and having a flora comparable with that of the Mesozoic in the northern hemisphere, though more varied. We have also the remarkable example of Australia, with a much richer flora than that of the early Mesozoic, yet inhabited only by non-placental mammals, like those of the Mesozoic.

The principal legacy that the Mesozoic woods have handed down to our time is in some beds of coal, locally important, but of far less extent than those of the Carboniferous period. Still, in America, the Richmond coalfield in Virginia is of this age, and so are the anthracite beds of the Queen Charlotte Islands, on the west coast of Canada, and the coal of Brora in Sutherlandshire. Valuable beds of coal, probably of this age, also exist in China, India, and South Africa; and jet, which is so extensively used for ornament, is principally derived from the carbonised remains of the old Mesozoic pines.

In the next chapter we have to study a revolution in vegetable life most striking and unique, in the advent of the forest-trees of strictly modern types.

NOTE TO CHAPTER V.

I APPEND to this chapter a table showing the plant-bearing series of the Cretaceous and Laramie of North America, from a paper in "Trans. R. S. C.," 1885, which see for further details:

(IN DESCENDING ORDER.)

Periods.	Floras and subfloras.	References.
Transition Eocene to Cretaceous.	Upper Laramie or Porcupine Hill. Fort Union group, U. S. territory.	Platanus beds of Souris River and Calgary. Report of Geol. Survey of Canada for 1879, and Memoir of 1885.
Upper Cretaceous (Danian and Senonian).	Middle Laramie or Willow Creek beds. Lower Laramie or St. Mary River. Fox Hill series .... Fort Pierre series........ Belly River.............. Coal measures of Nanaimo, B.C., probably here.	Lemna and Pistia beds of bad lands of 49th parallel, Red Deer River, &c., with lignites. Report 49th Parallel and Memoir of 1885. Marine. Marine. Sequoia and Brasenia beds of S. Saskatchewan, Belly River, &c., with lignites. Memoir of 1885. Memoir of 1883. Many dicotyledons, palms, &c.
Middle Cretaceous (Turonian and Cenomanian).	Dunvegan series of Peace River. Dakota group, U. S. Amboy clays, U. S. Mill Creek beds of Rocky Mountains.	Memoir of 1883. Many dicotyledons, cycads, &c. Dicotyledonous leaves, similar to Dakota group of the U. S. Memoir of 1885.
Lower Cretaceous (Neocomian, &c.).	Suskwa River beds and Queen Charlotte Island coal series. Intermediate beds of Rocky Mountains. Potomac series of Virginia. Kootanie series of Rocky Mountains.	Cycads, pines, a few dicotyledons. Report Geol. Survey. Memoir of 1885. Cycads, pines, and ferns. Memoir of 1885.

CHAPTER VI.

THE REIGN OF ANGIOSPERMS IN THE LATER CRETACEOUS AND KAINOZOIC.

IT is a remarkable fact in geological chronology that the culmination of the vegetable kingdom antedates that of the animal. The placental mammals, the highest group of the animal kingdom, are not known till the beginning of the Eocene Tertiary. The dicotyledonous Angiosperms, which correspond to them in the vegetable kingdom, occur far earlier—in the beginning of the Upper Cretaceous or close of the Lower Cretaceous. The reign of cycads and pines holds throughout the Lower Cretaceous, but at the close of that age there is a sudden incoming of the higher plants, and a proportionate decrease, more especially of the cycads.

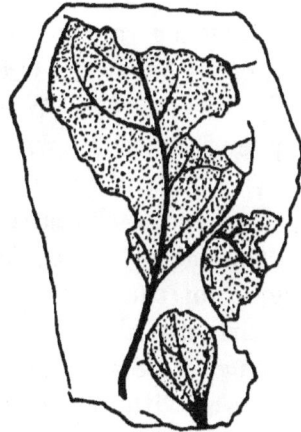

Fig. 68.—*Populus primæva*, Heer. Cretaceous, of Greenland. One of the oldest known Angiosperms.

I have already referred to the angiospermous wood supposed to be Devonian, but I fear to rest any conclusion on this isolated fact. Beyond this, the earliest indications of plants of this class have been found in the Lower Cretaceous. Many years ago Heer described and figured the leaves of a poplar (*Populus primæva*) from

the supposed Lower Cretaceous of Komé, in Greenland
(Fig. 68).   Two species, a *Sterculia* and a *Laurus* or
Salix, occur among fossils described by me in the upper
part of the Kootanie series of the Rocky Mountains, and
Fontaine has recently found in the Potomac group of
Virginia—believed to be of Neocomian age—several angiospermous species (*Sassafras, Menispermites, Sapindus,
Aralia, Populus,* &c.) mixed with a rich flora of cycads
and pines.   These are the early forerunners of the modern angiospermous flora ; but so far as known they do
not occur below the Cretaceous, and in its lower portions
only very rarely.   When, however, we ascend into the
Upper Cretaceous, whether of Europe or America, there
is a remarkable incoming of the higher plants, under
generic forms similar to those now existing.   This is, in
truth, the advent of the modern flora of the temperate
regions of the earth.   A very interesting tabular view of
its early distribution is given by Ward, in the "American
Journal of Science" for 1884, of which the following is a
synopsis, with slight emendations.   I may add that the
new discoveries made since 1884 would probably tend to
increase the proportionate number of dicotyledons in the
newer groups.

DICOTYLEDONOUS TREES IN THE CRETACEOUS.

Upper Senonian .... 179 species.
    (Fox Hill group of America.)

Lower Senonian .... 81 species.
    Upper white chalk of Europe ; Fort Pierre
      group of America ; coal-measures of Na-
      naimo ?

Turonian .... 20 species.
    Lower white chalk ; New Jersey marls ;
      Belly R. group.

Cenomanian .... 357 species.
    (Chalk-marl, greensand, and Gault, Niobrara
      and Dakota groups of America) ; Dun-
      vegan group of Canada ; Amboy clays of
      New Jersey.

Neocomian................................... 20 species.*

 (Lower greensand and Specton clay, Wealden
  and Hastings sands, Kootanie and Queen
  Charlotte groups of Canada.)

Thus we have a great and sudden inswarming of the higher plants of modern types at the close of the Lower Cretaceous. In relation to this, Saporta, one of the most enthusiastic of evolutionists, is struck by this phenomenon of the sudden appearance of so many forms, and some of them the most highly differentiated of dicotyledonous plants. The early stages of their evolution may, he thinks, have been obscure and as yet unobserved, or they may have taken place in some separate region, or mother country as yet undiscovered, or they may have been produced by a rapid and unusual multiplication of flower-haunting insects! Or it is even conceivable that the apparently sudden elevation of plants may have been due to causes still unknown. This last seems, indeed, the only certain inference in the case, since, as Saporta proceeds to say in conclusion : "Whatever hypothesis one may prefer, the fact of the rapid multiplication of dicotyledons, and of their simultaneous appearance in a great number of places in the northern hemisphere at the beginning of the Cenomanian epoch, cannot be disputed." †

The leaves described by Heer, from the Middle Cretaceous of Greenland, are those of a poplar (*P. primæva*). Those which I have described from a corresponding horizon in the Rocky Mountains are a *Sterculites* (*S. vetustula*), probably allied to the mallows, and an elongated leaf, *Laurophyllum* (*L. crassinerve*) (Fig. 69), which may, however, have belonged to a willow rather than a laurel. These are certainly older than the Dakota group

---

* Including an estimate of Fontaine's undescribed species.
† "Monde des Plantes," p. 197.

of the United States and the corresponding formations in Canada. On the eastern side of the American continent, in Virginia, the Potomac series is supposed to be of Lower Cretaceous age, and here Fontaine, as already stated, has found an abundant flora of cycads, conifers, and ferns, with a few angiospermous leaves, which have not yet been described.

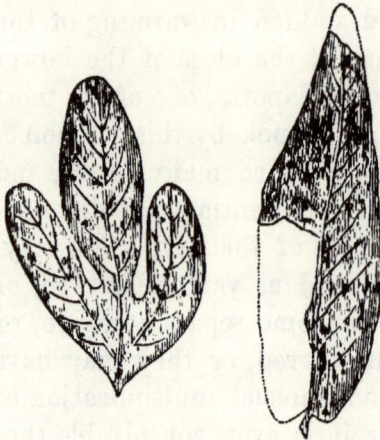

In the Canadian Rocky Mountains, a few hundreds of feet above the beds holding the before-mentioned species, are the shales of the Mill Creek series, rich in many species of dicotyledonous

FIG. 69.—*Stercalia* and *Laurophyllum* or *Salix*, the oldest Angiosperms known in the Cretaceous of Canada.

leaves, and corresponding in age with the Dakota group, whose fossils have been so well described, first by Heer and Capellini, and afterward by Lesquereux. We may take this Dakota group and the quader-sandstone of Germany as types of the plant-bearing Cenomanian, and may notice the forms occurring in them.

In the first place, we recognise here the successors of our old friends, the ferns and the pines, the latter represented by such genera as *Taxites, Sequoia, Glyptostrobus, Gingko,* and even *Pinus* itself. We also have a few cycads, but not so dominant as in the previous ages. The fan-palms are well represented, both in America and in the corresponding series in Europe, especially by the genus *Sabal,* which is the characteristic American type of fan-palm, and there is one genus which Saporta regards as intermediate between the fan-palms and the pinnately leaved species. There are also many fragments of stems

and leaves of carices and grasses, so that these plants, now
so important to the nourishment of man and his companion animals, were already represented.

Fig. 70.—Vegetation of Later Cretaceous.   Exogens and palms.   (After
Saporta.)

But the great feature of the time was its dicotyledonous forests, and I have only to enumerate the genera
supposed to be represented in order to show the richness
of the time in plants of this type.   It may be necessary
to explain here that the generic names used are mostly
based on leaves, and consequently cannot be held as being

absolutely certain, since we know that at present one
genus may have considerable variety in its leaves, and, on
the other hand, that plants of different genera may be
very much alike in their foliage.   There is, however, undoubtedly a likeness in plan or type of structure in leaves
of closely allied plants, and, therefore, if judiciously
studied, they can be determined with at least approxi-
mate certainty.*   More especially we can attain to much
certainty when the fruits as well as the leaves are found,
and when we can obtain specimens of the wood, showing
its structure.   Such corroboration is not wanting, though
unfortunately the leaves of trees are generally found
drifted away from the other organs once connected with
them.   In my own experience, however, I have often
found determinations of the leaves of trees confirmed by
the discovery of their fruits or of the structure of their
stems.   Thus, in the rich cretaceous plant-beds of the
Dunvegan series we have beech-nuts associated in the
same beds with leaves referred to *Fagus*.   In the Laramie
beds I determined many years ago nuts of the *Trapa*
or water-chestnut, and subsequently Lesquereux found,
in beds in the United States, leaves which he referred to
the same genus.   Later, I found in collections made on
the Red Deer River of Canada my fruits and Lesquereux's
leaves on the same slab.   The presence of trees of the
genera *Carya* and *Juglans* in the same formation was inferred from their leaves, and specimens have since been
obtained of silicified wood, with the microscopic structure
of the modern butternut.   Still we are willing to admit
that determinations from leaves alone are liable to doubt.

In the matter of names of fossil leaves, I sympathise
very strongly with Dr. Nathorst, of Stockholm, in his

---

* Great allowance has to be made for the variability of leaves of the
same species.   The modern hazel (*C. rostrata*) is a case in point.   Its
leaves, from different parts of the same plant, are so dissimilar in form
and size that they might readily be regarded as of different species.

objection to the use of modern generic names for mere
leaves, and would be quite content to adopt some non-
committal termination, as that of *"phyllum"* or *"ites"*
suggested by him. I feel, however, that almost as much
is taken for granted if a plant is called *Corylophyllum* or
Corylites, as if called *Corylus*. In either case a judgment
is expressed as to its affinities, which if wrong under the
one term is wrong under the other; and after so much has
been done by so many eminent botanists, it seems inex-
pedient to change the whole nomenclature for so small
and questionable an advantage. I wish it, however, to
be distinctly understood that plants catalogued on the
evidence of leaves alone are for the most part referred to
certain genera on grounds necessarily imperfect, and
their names are therefore subject to correction, as new
facts may be obtained.

The more noteworthy modern genera included in the
Dakota flora, as catalogued by Lesquereux, are the following : *Liquidambar*, the sweet-gum, is represented both in
America and Europe, the leaves resembling those of the
modern species, but with entire edges, which seems to be
a common peculiarity of Cretaceous foliage.* *Populus*
(poplar), as already stated, appears very early in Greenland, and continues with increasing number of species
throughout the Cretaceous and Tertiary. *Salix* (willow)
appears only a little later and continues. Of the family
Cupuliferæ we have *Fagus* (beech), *Quercus* (oak), and
Castanea (chestnut), which appear together in the Dakota
group and its equivalents. Fruits of some of the species
are known, and also wood showing structure. *Betula*

---

* With reference to this, something may be learned from the leaves
of modern trees. In these, young shoots have leaves often less toothed
and serrated than those of the adult tree. A remarkable instance is the
Populus grandidentatus of America, the young shoots of which have entire leaves, quite unlike except in venation those of the parent tree, and
having an aspect very similar to that of the Cretaceous poplars.

(birch) is represented by a few species, and specimens of
its peculiar bark are also common. *Alnus* (alder) appears in one species at least. The genus *Platanus* (Fig.
71), that of the plane-trees, represented at present by one

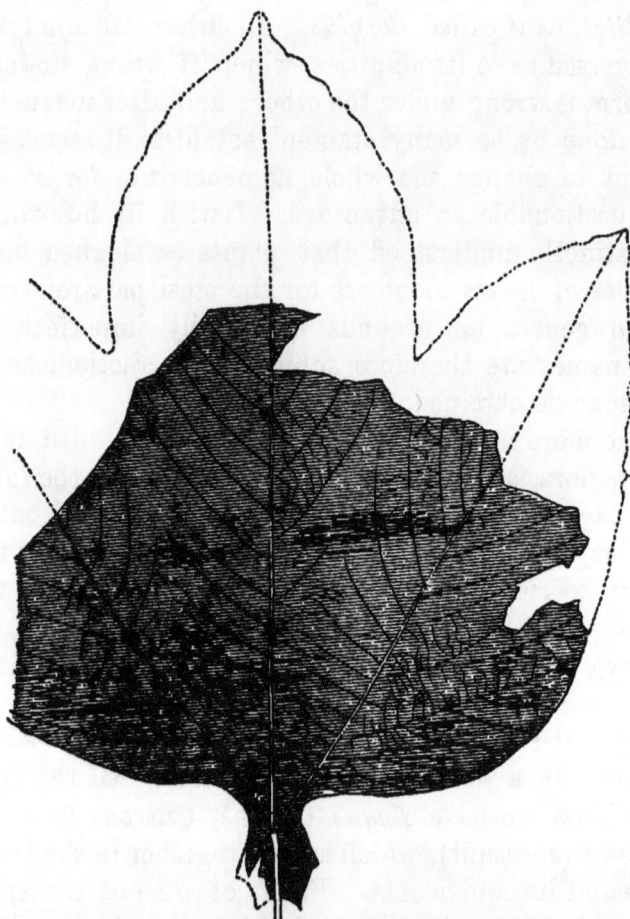

FIG. 71.—*Platanus nobilis*, Newberry, variety *basilobata*. Laramie.
Much reduced.

European and one American species, has several species
in the Cretaceous, though the plane-trees seem to culmi-
nate in the early part of the succeeding Eocene, where
there are several species with immense leaves. The large

leaves, known as *Credneria*, found in the Cenomanian of
Europe, and those called *Protophyllum* (Fig. 72) in
America, appear to be nearer to the plane-trees than to
any others, though representing an extinct type. The
laurels are represented in this age, and the American
genus *Sassafras*, which has now only one species, has not
one merely but several species in the Cretaceous. *Diospyros*, the persimmon-tree, was also a Cretaceous genus.

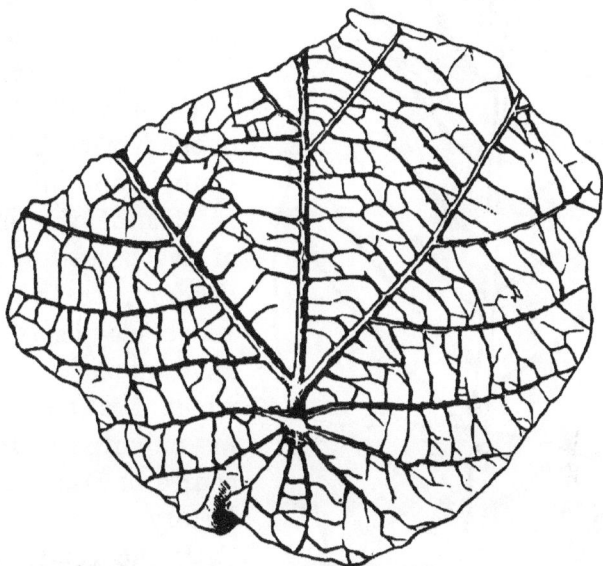

Fig. 72.—*Protophyllum boreale*, Dawson, reduced. Upper Cretaceous,
Canada.

The single species of the beautiful *Liriodendron*, or tulip-
tree, is a remnant of a genus which had several Cretaceous
species (Figs. 74, 75). The magnolias, still well represented in the American flora, were equally plentiful in the
Cretaceous (Fig. 73). The walnut family were well represented by species of *Juglans* (butternut) and *Carya*, or
hickory. In all, no less than forty-eight genera are present belonging to at least twenty-five families, running
through the whole range of the dicotyledonous exogens.
This is a remarkable result, indicating a sudden profusion

of forms of these plants of a very striking character.   It is further to be observed that some of the genera have many species in the Cretaceous and dwindle toward the modern.   In others the reverse is the case—they have expanded in modern times.   In a number there seems to have been little change.

Dr. Newberry has given, in the "Bulletin of the Torrey Botanical Club," an interesting *résumé* of the history of the beautiful *Liriodendron,* or tulip-tree, which may be taken as an example of a genus which has gone down in importance in the course of its geological history.

"The genus *Liriodendron,* as all botanists know, is represented in the present flora by a single species, 'the tulip-tree,' which is confined to eastern America, but grows over all the area lying between the Lakes and the Gulf, the Mississippi and the Atlantic.   It is a magnificent tree,   on the

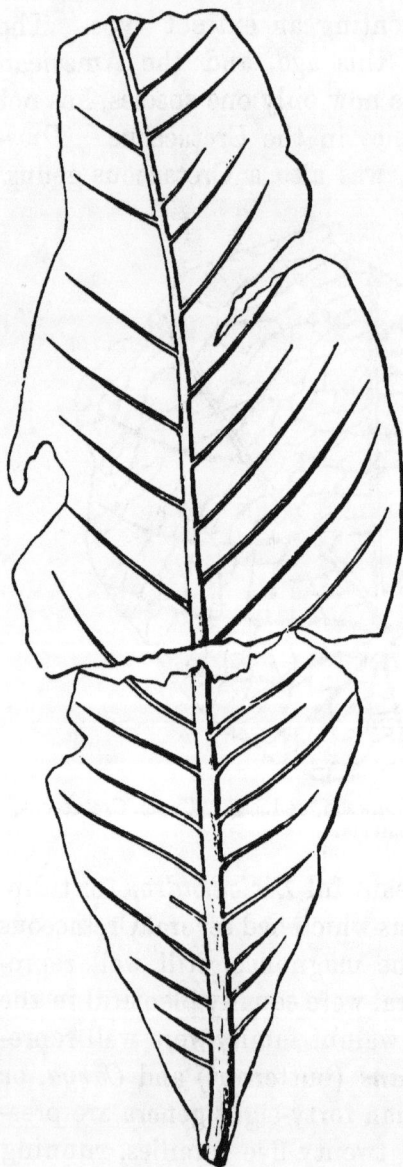

Fig. 73.—*Magnolia magnifica*, Dawson, reduced.   Upper Cretaceous, Canada.

whole, the finest in our forests. Its cylindrical trunk, sometimes ten feet in diameter, carries it beyond all its associates in size, while the beauty of its glossy, lyre-shaped leaves and tulip-like flowers is only surpassed by the flowers and foliage of its first cousin, *Magnolia grandiflora*. That a plant so splendid

Fig. 74.—*Liriodendron Meekii*, Heer. (After Lesquereux.)

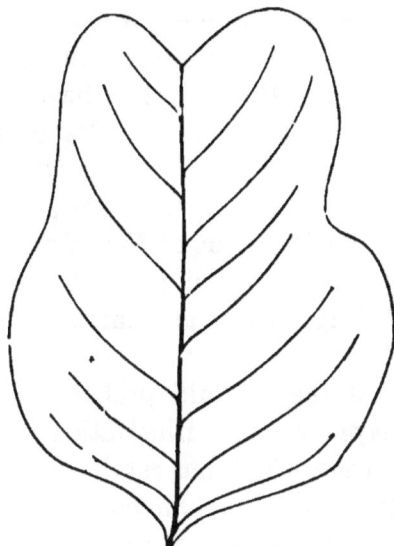

Fig. 75.—*Liriodendron primævum*, Newberry. (After Newberry.)

should stand quite alone in the vegetation of the present day excited the wonder of the earlier botanists, but the sassafras, the sweet-gum, and the great Sequoias of the far West afford similar examples of isolation, and the latter are still more striking illustrations of solitary grandeur." (Figs. 74 and 75.)

"Three species of *Liriodendron* are indicated by leaves found in the Amboy clays—Middle Cretaceous—of New Jersey, and others have been obtained from the Dakota group in the West, and from the Upper Cretaceous strata of Greenland. Though differing considerably among themselves in size and form, all these have the deep sinus of the upper extremity so characteristic of the genus, and the nervation is also essentially the same. Hence, we must conclude that the genus *Liriodendron*, now represented by a single species, was in the Cretaceous age much more largely developed, having many species, and those scattered throughout many lands. In the Tertiary age the genus continued to exist, but the species seem to have been reduced to one, which is hardly to be distinguished from that now living. In many parts of Europe leaves of the tulip-tree have been found, and it extended as far south as Italy. Its presence there was first made known by Unger, in his 'Synopsis,' page 232, and in his 'Genera et Species,' page 443, where he describes it under the name of *Liriodendron procaccinii*. The genus has also been noticed in Europe by Massalongo, Heer, and Ettingshausen, and three species have been distinguished. All these are, however, so much like the living species that they should. probably be united with it. We here have a striking illustration of the wide distribution of a species which has retained its characters both of fruit and leaf quite unchanged through long migrations and an enormous lapse of time.

" In Europe the tulip-tree, like many of its American associates, seems to have been destroyed by the cold of the Ice period, the Mediterranean cutting off its retreat, but in America it migrated southward over the southern extension of the continent and returned northward again with the amelioration of the climate."

Leaves of *Liriodendron* have been recognised in the Cretaceous of Greenland, though it is now a tree of the warm temperate region, and Lesquereux describes several species from the Dakota group. But the genus has not yet been recognised in the Laramie or in the Upper Cretaceous of British Columbia. In the paper above quoted, Newberry describes three new species from the Amboy clays, one of which he considers identical with a Greenland form referred by Heer to *L. Meeki* of the Dakota group. Thus, if all Lesquereux's species are to be accepted, the genus begins in the Middle Cretaceous with at least nine American species.

In New Jersey the Amboy clays are referred to the same age with the Dakota beds of the West. In these Dr. Newberry has found a rich flora, including many angiosperms. The following is condensed from a preliminary notice in the " Bulletin of the Torrey Botanical Club " : *

"The flora of the Amboy clays is closely related to that of the Dakota group—most of the genera and some of the species being identical—so that we may conclude they were nearly contemporaneous, though the absence in New Jersey of the Fort Benton and Niobrara groups of the upper Missouri and the apparent synchronism of the New Jersey marls and the Pierre group indicate that the Dakota is a little the older.

"At least one-third of the species of the Amboy clays seem to be identical with leaves found in the Upper Cretaceous clays of Greenland and Aachen (Aix la Chapelle), which not only indicates a chronological parallelism, but shows a remarkable and unexpected similarity in the vegetation of these widely separated countries in the middle and last half of the Cretaceous age. The botanical character of the flora of the Amboy clays will be seen from the following brief synopsis :

"*Algæ.*—A small and delicate form, allied to Chondrites.

"*Ferns.*—Twelve species, generally similar and in part identical with those described by Heer from the Cretaceous beds of Greenland, and referred to the genera *Dicksonia*, *Gleichenia*, and *Aspidium*.

"*Cycads.*—Two species, probably identical with the forms from Greenland described by Heer under the names of *Podozamites marginatus* and *P. tenuinervis*.

---

* March, 1886.

" *Conifers.*—Fourteen species, belonging to the genera *Moriconia, Brachyphyllum, Cunninghamites, Pinus, Sequoia,* and others referred by Heer to *Juniperus, Libocedrus, Frenelopsis, Thuya,* and *Dammara.* Of these, the most abundant and most interesting are *Moriconia cyclotoxon*—the most beautiful of conifers—and *Cunninghamites elegans,* both of which occur in the Cretaceous clays of Aachen, Prussia, and Patoot, Greenland. The *Brachyphyllum* was a large and strong species, with imbricated cones, eight inches in length.

" The angiosperms form about seventy species, which include three of *Magnolia,* four of *Liriodendron,* three or four of *Salix,* three of *Celastrophyllum* (of which one is identical with a Greenland species), one *Celastrus* (also found in Greenland), four or five *Aralias,* two *Sassafras,* one *Cinnamomum,* one *Hedera ;* with leaves that are apparently identical with those described by Heer as belonging to *Andromeda, Cissites, Cornus, Dewalquea, Diospyros, Eucalyptus, Ficus, Ilex, Juglans, Laurus, Menispermites, Myrica, Myrsine, Prunus, Rhamnus,* and others not yet determined.

" Some of the Aralias had palmately-lobed leaves, nearly a foot in diameter, and two of the tulip-trees (*Liriodendron*) had leaves quite as large as those of the living species. One of these had deeply lobed leaves, like those of the white oak. Of the other, the leaves resembled those of the recent tulip-tree, but were larger. Both had the peculiar emargination and the nervation of *Liriodendron.*

" Among the most interesting plants of the collection are fine species of *Bauhinia* and *Hymenæa.* Of these, the first is represented by a large number of leaves, some of which are six or seven inches in diameter. They are deeply bilobed, and have the peculiar and characteristic form and nervation of the leaves of this genus. *Bauhinia* is a leguminous genus allied to *Cercis,* and now in-

habits tropical and warm temperate climates in both
hemispheres. Only one species occurs in the United
States, *Bauhinia lunarioides,* Gray, found by Dr. Bige-
low on the Rio Grande.

" *Hymenœa* is another of the leguminosæ, and inhabits tropical America. A species of this genus has been
found in the Upper Cretaceous of France, but quite different from the one before us, in which the leaves are
much larger, and the leaflets are united in a common
petiole, which is winged ; this is a modification not found
in the living species, and one which brings it nearer to
Bauhinia.

" But the most surprising discovery yet made is that
of a number of quite large helianthoid flowers, which I
have called *Palœanthus.* These are three to four inches
in diameter, and exhibit a scaly involucre, enclosing what
much resembles a fleshy receptacle with achenia. From
the border of this radiate a number of ray florets, one to
two inches in length, which are persistent and must have
been scarious, like those of *Helichrysum.* Though these
flowers so much resemble those of the compositæ, we are
not yet warranted in asserting that such is certainly their
character. In the Jurassic rocks of Europe and India
some flowers not very unlike these have been found, which
have been named *Williamsonia,* and referred to cycads by
Carruthers. A similar fossil has been found in the Cretaceous rocks of Greenland, and named by Heer *Williamsonia cretacea,* but he questions the reference of the genus
to the Cycadeæ, and agrees with Nathorst in considering
all the species of *Williamsonia* as parasitic flowers, allied
to *Brugmansia* or *Rafflesia.* The Marquis of Saporta
regards them as monocotyledons, similar to *Pandanus.*
More specimens of the flowers now exhibited will perhaps
prove—what we can now only regard as probable—that
the Compositæ, like the *Leguminosæ, Magnoliaceæ, Ce-
lastraceæ,* and other highly organised plants, formed part

of the Cretaceous flora. No composite flowers have before been found in the fossil state, and, as these are among the most complex and specialised forms of florescence, it has been supposed that they belonged only to the recent epoch, where they were the result of a long series of formative changes."

The above presents some interesting new types not heretofore found in the Middle Cretaceous. More especially the occurrence of large flowers of the composite type presents a startling illustration of the early appearance of a very elevated and complex form. Great interest also attaches to these Amboy beds, as serving, with those of Aix and Greenland, to show that the margins of the Atlantic were occupied with a flora similar to that occurring at the same time in the interior plateau of North America and on the Pacific slope.

The beds at Aix-la-Chapelle are, however, probably somewhat newer than the Dakota or Amboy beds, and correspond more nearly in age with those of the Cretaceous coal-field of Vancouver Island, where there is a very rich Upper Cretaceous flora, which I have noticed in detail in the " Transactions of the Royal Society of Canada."* In these Upper Cretaceous beds there are fan-palms as far north at least as the latitude of 49°, indicating a very mild climate at this period. This inference is corroborated by the Upper Cretaceous flora of Atané and Patoot in Greenland, as described by Heer.

The dicotyledonous plants above referred to are trees and shrubs. Of the herbaceous exogens of the period we know less. Obviously their leaves are less likely to find their way into aqueous deposits than the leaves of trees. They are, besides, more perishable, and in densely wooded countries there are comparatively few herbaceous plants. I have examined the beds of mud deposited at the mouth

---

* Vol. ii., 1884.

of a woodland streamlet, and have found them stored with the fallen leaves of trees, but it was in vain to search for the leaves of herbaceous plants.

The climate of North America and Europe, represented by the Cenomanian vegetation, is not tropical but warm temperate ; but the flora was more uniform than at present, indicating a very equable climate and the possibility of temperate genera existing within the Arctic circle, and it would seem to have become warmer toward the close of the period.

The flora of the Cenomanian is separated in most countries from that of the Senonian, or uppermost Cretaceous, by a marine formation holding few plants. This depends on great movements of elevation and depression, to which we must refer in the sequel. In a few regions, however, as in the vicinity of the Peace River in Canada, there are plant-bearing beds which serve to bridge over the interval between the Early Cenomanian and the later Cretaceous.*

To this interval also would seem to belong the Belly River series of western Canada, which contains important beds of coal, but is closely associated with the marine Fort Pierre series. A very curious herbaceous

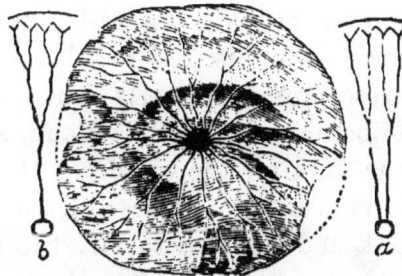

FIG. 76.—*Brasenia antiqua.* Upper Cretaceous, South Saskatchewan River. Natural size. *a, b,* Diagrams of venation, slightly enlarged.

plant of this group, which I have named *Brasenia antiqua,* occurs in the beds associated with one of the coals. It is a close ally of the modern *B. peltata,* an aquatic plant which occurs in British Columbia and in eastern

* See paper by the author in the " Transactions of the Royal Society of Canada," 1882.

America, and is also said to be found in Japan, Australia, and India, a width of distribution appropriate to so old a type (Fig. 76).

In so far as vegetable life is concerned, the transition from the Upper Cretaceous to the Tertiary or Kainozoic is easy, though in many parts of the world, and more especially in western Europe, there is a great gap in the deposits between the upper Chalk and the lowest Eocene. With reference to fossil plants, Schimper recognises in the Kainozoic, beginning with the oldest, five formations —Palæocene, Eocene, Oligocene, Miocene, and Pliocene. Throughout these a flora, similar to that of the Cretaceous on the one hand and the modern on the other, though with important local peculiarities, extends. There is evidence, however, of a gradual refrigeration, so that in the Pliocene the climates of the northern hemisphere were not markedly different from their present character.

In the first instance an important error was committed by palæobotanists, in referring to the Miocene many deposits really belonging to the Eocene. This arose from the early study of the rich plant-bearing Miocene beds of Switzerland, and from the similarity of the flora all the way from the Middle Cretaceous to the later Tertiary. The differences are now being worked out, and we owe to Mr. Starkie Gardner the credit of pointing these out in England, and to the Geological Survey of Canada that of collecting the material for exhibiting them in the more northern part of America.

In the great interior plain of America there rests on the Cretaceous a series of clays and sandstones with beds of lignite, some of them eighteen feet in thickness. This was formerly known as the lignitic or lignite Tertiary, but more recently as the Laramie series. These beds were deposited in fresh or brackish water, in an internal sea or group of lakes and swamps, when the continent was lower than at present. They have been st·idied both in the United States* and Canada ; and, though their flora was originally referred by mistake to the Miocene, it is now known to be Eocene or Palæocene, or even in part a transition group between the latter and the Cretaceous.   The following remarks, taken chiefly from recent papers by the author,† will serve to illustrate this :

On the geological map of Canada the Laramie series, formerly known as the lignitic or lignite Tertiary, occurs, with the exception of a few outliers, in two large areas west of the 100th meridian, and separated from each other by a tract of older Cretaceous rocks, over which the Laramie beds may have extended, before the later denudation of the region.

The most eastern of these areas, that of the Souris River and Wood Mountain, extends for some distance along the United States boundary, between the 102d and 109th meridians, and reaches northward to about thirty miles south of the "elbow" of the South Saskatchewan River, which is on the parallel of 51° north.   In this area the lowest beds of the Laramie are seen to rest on those of the Fox Hill group of the Upper Cretaceous, and at one point on the west they are overlaid by beds of Miocene Tertiary age, observed by Mr. McConnell, of the Geological Survey, in· the Cypress Hills, and referred by Cope, on the evidence of mammalian remains, to the White River division of the United States geologists, which is regarded by them as Lower Miocene.‡   The age of the Laramie beds is thus stratigraphically determined to be between the Fox Hill Cretaceous and the Lower

---

* See more especially the elaborate and valuable reports by Lesquereux and Newberry, and a recent memoir by Ward on "Types of the Laramie Flora," "Bulletins of the United States Geological Survey," 1887.

† "Transactions of the Royal Society of Canada," 1886–'87.

‡ "Report of the Geological Survey of Canada," 1885.

Miocene.   They are also undoubtedly continuous with the Fort Union group of the United States geologists on the other side of the international boundary, and they contain similar fossil plants.   They are divisible into two groups—a lower, mostly argillaceous, and to which the name of "Bad Lands beds" may be given, from the "bad lands" of Wood Mountain, where they are well exposed, and an upper, partly arenaceous member, which may be named the Souris River or Porcupine Creek division. In the lower division are found reptilian remains of Upper Cretaceous type, with some fish remains more nearly akin to those of the Eocene.* Neither division has as yet afforded mammalian remains.

The western area is of still larger dimensions, and extends along the eastern base of the Rocky Mountains from the United States boundary to about the 55th parallel of latitude, and stretches eastward to the 111th meridian. In this area, and more especially in its southern part, the officers of the Geological Survey of Canada have recognised three divisions, as follows : (1) The Lower Laramie or St. Mary River series, corresponding in its character and fossils to the Lower or Bad Lands division of the other area.   (2) A middle division, the Willow Creek beds, consisting of clays, mostly reddish, and not recognised in the other area.   (3) The Upper Laramie or Porcupine Hills division, corresponding in fossils, and to some extent in mineral character, to the Souris River beds of the eastern area.

The fossil plants collected by Dr. G. M. Dawson in the eastern area were noticed by the author in an appendix to Dr. Dawson's report on the 49th parallel, in 1875, and a collection subsequently made by Dr. Selwyn was described in the "Report of the Geological Survey of Canada" for 1879–'80.   Those of the western area, and

---

* Cope, in Dr. G. M. Dawson's "Report on the 49th Parallel."

especially collections made by myself near Calgary in 1883, and by officers of the Geological Survey in 1884, have been described in the "Transactions of the Royal Society of Canada," vols. iii. and iv.

In studying these fossil plants, I have found that there is a close correspondence between those of the Lower and Upper Laramie in the two areas above referred to respectively, and that the flora of the Lower Laramie is somewhat distinct from that of the Upper, the former being especially rich in certain aquatic plants, and the latter much more copious on the whole, and much more rich in remains of forest-trees. This is, however, possibly an effect rather of local conditions than of any considerable change in the flora, since some Upper Laramie forms recur as low as the Belly River series of the Cretaceous, which is believed on stratigraphical grounds to be considerably older than the Lower Laramie.

With reference to the correlation of these beds with those of the United States, some difficulty has arisen from the tendency of palæobotanists to refer the plants of the Upper Laramie to the Miocene age, although in the reports of Mr. Clarence King, the late director of the United States Geological Survey, these beds are classed, on the evidence of stratigraphy and animal fossils, as Upper Cretaceous. More recently, however, and partly perhaps in consequence of the views maintained by the writer since 1875, some change of opinion has occurred, and Dr. Newberry and Mr. Lesquereux seem now inclined to admit that what in Canada we recognise as Upper Laramie is really Eocene, and the Lower Laramie either Cretaceous or a transition group between this and the Eocene. In a recent paper * Dr. Newberry gives a comparative table, in which he correlates the Lower

---

* Newberry, "Transactions of the New York Academy," February, 1886.

Laramie with the Upper Cretaceous of Vancouver Island and the Faxoe and Maestricht beds of Europe, while he regards the Upper Laramie as equivalent to European Eocene. Except in so far as the equivalence of the Lower Laramie and Vancouver Island beds is concerned, this corresponds very nearly with the conclusions of the writer in a paper published last year *—namely, that we must either regard the Laramie as a transition Cretaceo-Eocene group, or must institute our line of separation in the Willow Creek or Middle Laramie division, which has, however, as yet afforded no fossil plants. I doubt, however, the equivalence of the Vancouver beds and the Lower Laramie, except perhaps in so far as the upper member of the former is concerned. I have also to observe that in the latest report of Mr. Lesquereux he still seems to retain in the Miocene certain formations in the West, which from their fossil plants I should be inclined to regard as Eocene.†

Two ferns occurring in these beds are remarkable as evidence of the persistence of species, and of the peculiarities of their ancient and modern distribution. *Onoclea sensibilis*, the very common sensitive fern of eastern America, is extremely abundant in the Laramie beds over a great area in the West. Mr. Starkie Gardner and Dr. Newberry have also shown that it is identical with the *Filicites Hebridicus* of Forbes, from the early Eocene beds of the Island of Mull, in Scotland. Thus we have a species once common to Europe and America, but now restricted to the latter, and which has continued to exist over all the vast ages between the Cretaceous and the present day. In the Laramie beds I have found asso-

---

* "Transactions of the Royal Society of Canada," vol. ii.

† While these sheets were going through the press I received a very valuable report of Mr. Lester F. Ward upon the Laramie of the United States. I have merely had time to glance at this report, but can see that the views of the author agree closely with those above expressed.

ciated with this species another and more delicate fern, the modern *Davallia* (*Stenloma*) *tenuifolia*, but this, unlike its companion, no longer occurs in America, but is found in the mountains of Asia. This is a curious illustration of the fact that frail and delicate plants may be more ancient than the mountains or plains on which they live.

There are also some very interesting and curious facts in connection with the conifers of the Laramie. One of the most common of these is a *Thuja* or arbor vitæ (the so-called "cedar" of Canada). The Laramie species has been named *T. interrupta* by Newberry, but it approaches very closely in its foliage to *T. occidentalis*, of eastern Canada, while its fruit resembles that of the western species, *T. gigantea*.

Still more remarkable are the Sequoias to which we have already referred, but which in the Laramie age seem to have been spread over nearly all North America. The fossil species are of two types, representing respectively the modern *S. gigantea* and *S. sempervirens*, and their wood, as well as that of Thuja, is found in great abundance in the lignites; and also in the form of silicified trunks, and corresponds with that of the recent species. The Laramie contains also conifers of the genera *Glyptostrobus, Taxodium*, and *Taxus ;* and the genus *Salisburia* or gingko—so characteristic of the Jurassic and Cretaceous—is still represented in America as well as in Europe in the early Eocene.

We have no palms in the Canadian or Scottish Palæocene, though I believe they are found further south. The dicotyledonous trees are richly represented. Perhaps the most conspicuous were three species of *Platanus*, the leaves of which sometimes fill the sandstones, and one of which, *P. nobilis*, Newberry, sometimes attains the gigantic size of a foot or more in diameter of its blade. The hazels are represented by a large-leaved species, *C.*

Macquarii, and by leaves not distinguishable from those of the modern American species, *C. Americana* and *C. rostrata.* There are also chestnuts and oaks. But the poplars and willows are specially abundant, being represented by no less than six species, and it would seem that all the modern types of poplar, as indicated by the forms and venation of the leaves, existed already in the Laramie, and most of them even in the Upper Cretaceous. *Sassafras* is represented by two species, and the beautiful group of *Viburnum,* to which the modern tree-cranberry belongs, has several fine species, of some of which both leaves and berries have been found. The hickories and butternuts are also present, the horse-chestnut, the *Catalpa* and *Sapindus,* and some curious leaves which seem to indicate the presence of the modern genus *Symphorocarpus,* the snow-berry tribe.

The above may suffice to give an idea of the flora of the older Eocene in North America, and I may refer for details to the works of Newberry, Lesquereux, and Ward, already cited. I must now add that the so-called Miocene of Atanekerdluk, Greenland, is really of the same age, as also the "Miocene" of Mull, in Scotland, of Antrim, in Ireland, and of Bovey Tracey, in the south of England, and the Gelinden, or "Heersian" beds, of Belgium, described by Saporta. In comparing the American specimens with the descriptions given by Gardner of the leaf-beds at Ardtown, in Mull, we find, as already stated, *Onoclea sensibilis,* common to both. The species of *Sequoia, Gingko, Taxus,* and *Glyptostrobus* are also identical or closely allied, and so are many of the dicotyledonous leaves. For example, *Platanoides Hebridicus* is very near to *P. nobilis,* and *Corylus Macquarrii* is common to both formations, as well as *Populus Arctica* and *P. Richardsoni.* I may add that ever since 1875-'76, when I first studied the Laramie plants, I have maintained their identity with those of the Fort Union group

of the United States, and of the so-called Miocene of
McKenzie River and Greenland, and that the whole are
Paleocene ; and this conclusion has now been confirmed
by the researches of Gardner in England, and by the discovery of true Lower Miocene beds in the Canadian northwest, overlying the Laramie or lignite series.

In a bulletin of the United States Geological Survey (1886), Dr. White has established in the West the
continuous stratigraphical succession of the Laramie and
the Wahsatch Eocene, thus placing the Laramie conformably below the Lower Eocene of that region.   Cope
has also described as the Puerta group a series of beds
holding vertebrate fossils, and forming a transition from
the Laramie to the Wahsatch.   White also testifies that a
number of fresh-water mollusks are common to the Wahsatch and the Laramie.   This finally settles the position
of the Laramie so far as the United States geologists are
concerned, and shows that the flora is to be regarded as
Eocene if not Upper Cretaceous, in harmony with what
has been all along maintained in Canada.   An important
résumé of the flora has just been issued by Ward in the
bulletins of the United States Geological Survey (1887).

Before leaving this part of the subject, I would depre-
cate the remark, which I see occasionally made, that fossil
plants are of little value in determining geological horizons in the Cretaceous and Tertiary.   I admit that in
these periods some allowance must be made for local
differences of station, and also that there is a generic
sameness in the flora of the northern hemisphere, from
the Cenomanian to the modern, yet these local differences and general similarity are not of a nature to invalidate inferences as to age.   No doubt, so long as
palæobotanists seemed obliged, in deference to authority,
and to the results of investigations limited to a few European localities, to group together, without distinction,
all the floras of the later Cretaceous and earlier Tertiary,

irrespective of stratigraphical considerations, the subject lost its geological importance. But, when a good series has been obtained in any one region of some extent, the case becomes different. Though there is still much imperfection in our knowledge of the Cretaceous and Tertiary floras of Canada, I think the work already done is sufficient to enable any competent observer to distinguish by their fossil plants the Lower, Middle, and Upper Cretaceous, and the latter from the Tertiary ; and, with the aid of the work already done by Lesquereux and Newberry in the United States, to refer approximately to its true geological position any group of plants from beds of unknown age in the West.

An important consequence arising from the above statements is that the period of warm climate which enabled a temperate flora to exist in Greenland was that of the later Cretaceous and early Eocene rather than, as usually stated, the Miocene. It is also a question admitting of discussion whether the Eocene flora of latitudes so different as those of Greenland, Mackenzie River, northwest Canada, and the United States, were strictly contemporaneous, or successive within a long geological period in which climatal changes were gradually proceeding. The latter statement must apply at least to the beginning and close of the period ; but the plants themselves have something to say in favour of contemporaneity. The flora of the Laramie is not a tropical but a temperate flora, showing no doubt that a much more equable climate prevailed in the more northern parts of America than at present. But this equability of climate implies the possibility of a great geographical range on the part of plants. Thus it is quite possible and indeed highly probable that in the Laramie age a somewhat uniform flora extended from the Arctic seas through the great central plateau of America far to the south, and in like manner along the western coast of

Europe. It is also to be observed that, as Gardner points out, there are some differences indicating a diversity of climate between Greenland and England, and even between Scotland and Ireland and the south of England, and we have similar differences, though not strongly marked, between the Laramie of northern Canada and that of the United States. When all our beds of this age from the Arctic sea to the 49th parallel have been ransacked for plants, and when the palæobotanists of the United States shall have succeeded in unravelling the confusion which now exists between their Laramie and the Middle Tertiary, the geologist of the future will be able to restore with much certainty the distribution of the vast forests which in the early Eocene covered the now bare plains of interior America. Further, since the break which in western Europe separates the flora of the Cretaceous from that of the Eocene does not exist in America, it will then be possible to trace the succession from the Mesozoic flora of the Trias and of the Queen Charlotte Islands and Kootanie series of the Lower Cretaceous up to the close of the Eocene; and to determine, for America at least, the manner and conditions under which the angiospermous flora of the later Cretaceous succeeded to the pines and cycads which characterised the beginning of the Cretaceous period. In so far as Europe is concerned, this may be more difficult, since the want of continuity of land from north to south seems there to have been fatal to the continuance of some plants during changes of climate, and there were also apparently in the Kainozoic period invasions at certain times of species from the south and east, which did not occur to the same extent in America.

In recent reports on the Tertiary floras of Australia and New Zealand,* Ettingshausen holds that the flora of

---

* "Geological Magazine," August, 1887.

the Tertiary, as a whole, was of a generalised character; forms now confined to the southern and northern hemispheres respectively being then common to both. It would thus seem that the present geographical diversities must have largely arisen from the great changes in climate and distribution of land and water in the later Tertiary.

The length of our discussion of the early angiospermous flora does not permit us to trace it in detail through the Miocene and Pliocene, but we may notice the connection through these in the next chapter, and may refer to the magnificent publications of Heer and Lesquereux on the Tertiary floras of Europe and America respectively.

CHAPTER VII.

It may be well to begin this chapter with a sketch of the general physical and geological conditions of the period which was characterised by the advent and culmination of the dicotyledonous trees.

In the Jurassic and earliest Cretaceous periods the prevalence, over the whole of the northern hemisphere and for a long time, of a monotonous assemblage of gymnospermous and acrogenous plants, implies a uniform and mild climate, and facility for intercommunication in the north. Toward the end of the Jurassic and beginning of the Cretaceous, the land of the northern hemisphere was assuming greater dimensions, and the climate probably becoming a little less uniform. Before the close of the Lower Cretaceous period the dicotyledonous flora seems to have been introduced, under geographical conditions which permitted a warm temperate climate to extend as far north as Greenland.

In the Cenomanian or Middle Cretaceous age we find the northern hemisphere tenanted with dicotyledonous trees closely allied to those of modern times, though still indicating a climate much warmer than that which at present prevails. In this age, extensive but gradual submergence of land is indicated by the prevalence of chalk and marine limestones over the surface of both continents; but a circumpolar belt seems to have been maintained, protecting the Atlantic and Pacific basins from floating ice, and permitting a temperate flora of great richness to prevail far to the north, and especially along the southern margins and extensions of the circumpolar land. These seem to have been the physical conditions which terminated the existence of the old Mesozoic flora and introduced that of the Middle Cretaceous.

As time advanced the quantity of land gradually increased, and the extension of new plains along the older ridges of land was coincident with the deposition of the great Laramie series, and with the origination of its peculiar flora, which indicates a mild climate and considerable variety of station in mountain, plain, and swamp, as well as in great sheets of shallow and weedy fresh water.

In the Eocene and Miocene periods, the continents gradually assumed their present form, and the vegetation became still more modern in aspect. In that period of the Eocene, however, in which the great nummulitic limestones were deposited, a submergence of land occurred on the eastern continent which must have assimilated its physical conditions to those of the Middle Cretaceous. This great change, affecting materially the flora of Europe, was not equally great in America, which also by the north and south extension of its mountain-chains permitted movements of migration not possible in the Old World. From the Eocene downward, the remains of land-animals and plants are found chiefly in lake-basins occupying the existing depressions of the land, though more extensive than those now remaining. It must also be borne in mind that the great foldings and fractures of the crust of the earth which occurred at the close of the Eocene, and to which the final elevation of such ranges as the Alps and the Rocky Mountains belongs, permanently modified and moulded the forms of the continents.

These statements raise, however, questions as to the precise equivalence in time of similar floras found in different latitudes. However equable the climate, there must have been some appreciable difference in proceeding from north to south. If, therefore, as seems in every way probable, the new species of plants originated on the Arctic land and spread themselves southward, this latter process would occur most naturally in times of gradual refrigeration or of the access of a more extreme climate—that is, in times of the elevation of land in the temperate latitudes, or, conversely, of local depression of land in the Arctic, leading to invasions of northern ice. Hence, the times of the prevalence of particular types of plants in the far north would precede those of their extension to the south, and a flora found fossil in Greenland might be supposed to be somewhat older than a similar flora when found farther south. It would seem, however, that the time required for the extension of a new flora to its extreme geographical limit is so small, in comparison with the duration of an entire geological period, that, practically, this difference is of little moment, or at least does not amount to antedating the Arctic flora of a particular type by a whole period, but only by a fraction of such period.

It does not appear that, during the whole of the Cretaceous and Eocene periods, there is any evidence of such refrigeration as seriously to interfere with the flora, but perhaps the times of most considerable warmth are those of the Dunvegan group in the Middle Cretaceous, and those of the later Laramie and oldest Eocene.

It would appear that no cause for the mild temperature of the Cretaceous needs to be invoked, other than those mutations of land and water which the geological deposits themselves indicate. A condition, for example, of the Atlantic basin in which the high land of Greenland should be reduced in elevation, and at the same time the northern inlets of the Atlantic closed against the invasion of Arctic ice, would at once restore climatic conditions allowing of the growth of a temperate flora in Greenland. As Dr. Brown has shown,* and as I have elsewhere argued, the absence of light in the Arctic winter is no disadvantage, since, during the winter, the growth of deciduous trees is in any case suspended ; while the constant continuance of light in the summer is, on the contrary, a very great stimulus and advantage.

It is a remarkable phenomenon in the history of genera of plants in the later Mesozoic and Tertiary, that the older genera appear at once in a great number of specific types, which become reduced as well as limited in range down to the modern. This is, no doubt, connected with the greater differentiation of local conditions in the modern ; but it indicates also a law of rapid multiplication of species in the early life of genera. The distribution of the species of *Salisburia, Sequoia, Platanus, Sassafras, Liriodendron, Magnolia,* and many other genera, affords remarkable proofs of this.

Gray, Saporta, Heer, Newberry, Lesquereux, and Starkie Gardner have all ably discussed these points ; but the continual increase of our knowledge of the several floras, and the removal of error as to the dates of their appearance, must greatly conduce to clearer and more definite ideas. In particular, the prevailing opinion that the Miocene was the period of the greatest extension of warmth and of a temperate flora into the Arctic, must be abandoned in favour of the later Cretaceous and Eocene ; and, if I mistake not, this will be found to accord better with the evidence of general geology and of animal fossils.

In these various revolutions of the later Cretaceous and Kainozoic periods, America, as Dr. Gray has well pointed out, has had the advantage of a continuous stretch of high land from north to south, affording a more sure

---

* " Florula Discoana."

refuge to plants in times of submergence, and means of escape to the south in times of refrigeration. Hence, the greater continuity of American vegetation and the survival of genera like *Sequoia* and *Liriodendron*, which have perished in the Old World. Still, there are some exceptions to this, for the gingko-tree is a case of survival in Asia of a type once plentiful in America, but now extinct there. Eastern Asia has had, however, some considerable share of the same advantage possessed by America, with the addition, referred to by Gray, of a better and more insular climate.

But our survey of these physical conditions can not be considered complete till we shall have considered the great Glacial age of the Pleistocene. It is certain that throughout the later Miocene and Pliocene the area of land in the northern hemisphere was increasing, and the large and varied continents were tenanted by the noblest vegetation and the grandest forms of mammalian life that the earth has ever witnessed. As the Pliocene drew to a close, a gradual diminution of warmth came on, and more especially a less equable climate, and this was accompanied with a subsidence of the land in the temperate regions and with changes of the warm ocean-currents. Thus gradually the summers became cooler and the winters longer and more severe, the hill-tops became covered with permanent snows, glaciers ploughed their way downward into the plains, and masses and fields of floating ice cooled the seas. In these circumstances the richer and more delicate forms of vegetation must have been chilled to death or obliged to remove farther south, and in many extensive regions, hemmed in by the advance of the sea on the one hand and land-ice on the other, they must have altogether perished.

Yet even in this time vegetation was not altogether extinct. Along the Gulf of Mexico in America, and in the Mediterranean basin in Europe, there were still some remains of a moderate climate and certain boreal and arctic forms moving southward continued to exist here and there in somewhat high latitudes, just as similar plants now thrive in Grinnell Land within sight of the snows of the Greenland mountains. A remarkable summary of some of these facts as they relate to England was given by an eminent English botanist, Mr. Carruthers, in his address as President of the Biological Section of the British Association at Birmingham in 1886. At Cromer, on the coast of Norfolk, the celebrated forest-bed of newer Pliocene age, and containing the remains of a copious mammalian fauna, holds also remains of plants in a state admitting of determination. These have been collected by Mr. Reid, of the Geological Survey, and were reported on by Carruthers, who states that they represent a somewhat colder temperature than that of the present day. I quote the following details from the address.

With reference to the plants of the forest-bed or newer Pliocene he remarks as follows :

"Only one species (*Trapa natans*, Willd.) has disappeared from our islands. Its fruits, which Mr. Reid found abundantly in one locality, agree with those of the plants found until recently in the lakes of Sweden. Four species (*Prunus speciosa*, L., *Œnanthe Tichenalii*, Sm., *Potamogeton pterophyllus*, Sch., and *Pinus abies*, L.) are found at present only in Europe, and a fifth (*Potamogeton trichoides*, Cham.) extends also to North America ; two species (*Peucedanum palustre*, Moench, and *Pinus sylvestris*, L.) are found also in Siberia, while six more (*Sanguisorba officinalis*, L., *Rubus fruticosus*, L., *Cornus sanguinea*, L., *Euphorbia amygdaloides*, L., *Quercus robur*, L., and *Potamogeton crispus*, L.) extend into western Asia, and two (*Fagus sylvatica*, L., and *Alnus glutinosa*, L.) are included in the Japanese flora. Seven species, while found with the others, enter also into the Mediterranean flora, extending to North Africa : these are *Thalictrum minus*, L., *Thalictrum flavum*, L., *Ranunculus repens*, L., *Stellaria aquatica*, Scop., *Corylus avellana*, L., *Yannichellia palustris*, L., and *Cladium mariscus*, Br.   With a similar distribution in the Old World, eight species (*Bidens tripartita*, L., *Myosotis cæspitosa*, Schultz, *Suæda maritima*, Dum., *Ceratophyllum demersum*, L., *Sparganium ramosum*, Huds., *Potamogeton pectinatus*, L., *Carex paludosa*, Good., and *Osmunda regalis*, L.) are found also in North America.   Of the remainder, ten species (*Nuphar luteum*, Sm., *Menyanthes trifoliata*, L., *Stachys palustris*, L., *Rumex maritimus*, L., *Rumex acetosella*, L., *Betula alba*, L., *Scirpus pauciflorus*, Lightf., *Taxus baccata*, L., and *Isoetes lacustris*, L.) extend round the north temperate zone, while three (*Lycopus europæus*, L., *Alisma plantago*, L., and *Phragmites communis*, Trin.), having the same distribution in the north, are found also in Australia, and one (*Hippuris vulgaris*, L.) in the south of South America. The list is completed by *Ranunculus aquatilis*, L., distributed over all the temperate regions of the globe, and *Scirpus lacustris*, L., which is found in many tropical regions as well."

He remarks that these plants, while including species now very widely scattered, present no appreciable change of characters.

Above this bed are glacial clays, which hold other species indicating an extremely cold climate.   They are few in number, only *Salix polaris*, a thoroughly arctic species, and its ally, *S. cinerea*, L., and a moss, *Hypnum turgescens*, Schimp., no longer found in Britain, but an Alpine and arctic species.   This bed belongs to the beginning of the Glacial period, the deposits of which have as yet afforded no plants in England.   But plants occur in post-glacial and upper-glacial beds in different parts of England, to which Carruthers thus refers :

"The period of great cold, during which arctic ice

Q

extended far into temperate regions, was not favorable to vegetable life. But in some localities we have stratified clays with plant-remains later than the Glacial epoch, yet indicating that the great cold had not then entirely disappeared. In the lacustrine beds at Holderness is found a small birch (*Betula nana*, L.), now limited in Great Britain to some of the mountains of Scotland, but found in the arctic regions of the Old and New World and on Alpine districts in Europe, and with it *Prunus padus*, L., *Quercus robur*, L., *Corylus avellana*, L., *Alnus glutinosa*, L., and *Pinus sylvestris*, L. In the white clay-beds at Bovey Tracey of the same age there occur the leaves of *Arctostaphylos uva-ursi*, L., three species of willow, viz., *Salix cinerea*, L., *S. myrtilloides*, L., and *S. polaris*, Wahl., and in addition to our Alpine *Betula nana*, L., the more familiar *B. alba*, L. Two of these plants have been lost to our flora from the change of climate that has taken place, viz., *Salix myrtilloides*, L., and *S. polaris*, Wahl.; and *Betula nana*, L., has retreated to the mountains of Scotland. Three others (*Dryas octopetala*, L., *Arctostaphylos uva-ursi*, L., and *Salix herbacea*, L.) have withdrawn to the mountains of northern England, Wales, and Scotland, while the remainder are still found scattered over the country. Notwithstanding the diverse physical conditions to which these plants have been subjected, the remains preserved in these beds present no characters by which they can be distinguished from the living representatives of the species."

One of the instances referred to is very striking. At Bovey Tracey the arctic beds rest directly on those holding the rich, warm temperate flora of the Eocene; so that here we have the evidence of fossil plants to show the change from the climate of the Eocene to that of arctic lands, and the modern vegetation to indicate the return of a warm temperature.

In Canada, in the Pleistocene beds known as the Leda clays, intervening between the lower boulder clay and the Saxicava sand, which also holds boulders, there are beds holding fossil plants, in some places intermixed with sea-shells and bones of marine fishes, showing that they were drifted into the sea at a time of submergence. These remains are boreal rather than arctic in character, and with the remains of drift-wood often found in the boulder deposits serve to indicate that there were at all times oases of hardy life in the glacial deserts, just as we find these in polar lands at the present day. I condense from a paper on these plants * the following facts, with a few additional notes :

The importance of all information bearing on the temperature of the Post - pliocene period invests with much interest the study of the land-plants preserved in deposits of this age. Unfortunately, these are few in number, and often not well preserved. In Canada, though fragments of the woody parts of plants occasionally occur in the marine clays and sands, there is only one locality which has afforded any considerable quantity of remains of their more perishable parts. This is the well-known deposit of Leda clay at Green's Creek, on the Ottawa, celebrated for the perfection in which the skeletons of the capelin and other fishes are preserved in the calcareous nodules imbedded in the clay. In similar nodules, contained apparently in a layer somewhat lower than that holding the ichthyolites, remains of land-plants are somewhat abundant, and, from their association with shells of *Leda glacialis*, seem to have been washed down from the land into deep water. The circumstances would seem to have been not dissimilar from those at present existing in the northeast arm of Gaspé Basin, where I have dredged from mud now being deposited in deep water, living

* "Canadian Naturalist," 1866.

specimens of *Leda limatula*, mixed with remains of land-plants.

The following are the species of plants recognised in these nodules :

1. *Drosera rotundifolia*, Linn.   In a calcareous nodule from Green's Creek, the leaf only preserved.   This plant is common in bogs in Canada, Nova Scotia, and Newfoundland, and thence, according to Hooker, to the Arctic circle.   It is also European.

2. *Acer spicatum*, Lamx.   (*Acer montanum*, Aiton.) Leaf in a nodule from Green's Creek.   Found in Nova Scotia and Canada, also at Lake Winnipeg, according to Richardson.

3. *Potentilla Canadensis*, Linn.   In nodules from Green's Creek ; leaves only preserved.   I have had some difficulty in determining these, but believe they must be referred to the species above named, or to *P. simplex*, Michx., supposed by Hooker and Gray to be a variety.   It occurs in Canada and New England, but I have no information as to its range northward.

Fig. 77.—*Gaylussaccia resinosa*.   Pleistocene, Canada.

4. *Gaylussaccia resinosa*, Torrey and Gray.   Leaf in nodule at Green's Creek.   Abundant in New England and in Canada, also on Lake Huron and the Saskatchewan, according to Richardson (Fig. 77).

5. *Populus balsamifera*, Linn.   Leaves and branches in nodules at Green's Creek.   This is by much the most common species, and its leaves are of small size, as if from trees growing in cold and exposed situations.   The species is North American and Asiatic, and abounds in New England and Canada.   It extends to the Arctic circle, and is

abundant on the shores of the Great Slave Lake and on the McKenzie River, and according to Richardson constitutes much of the drift timber of the Arctic coast (Fig. 78).

6. *Thuja occidentalis*, Linn.   Trunks and branches in the Leda clay at Montreal.   This tree occurs in New England and Canada, and extends northward into the

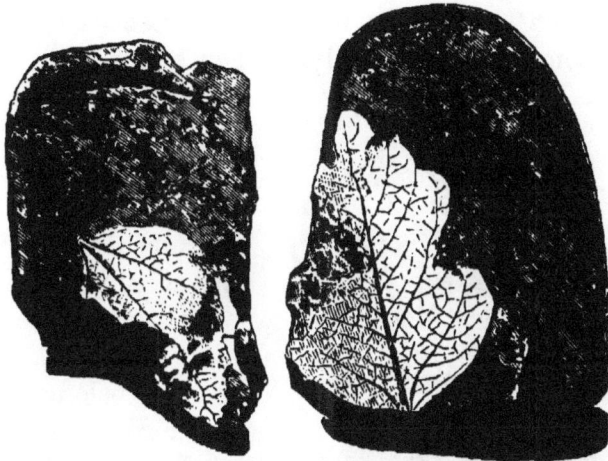

Fɪɢ. 78.—*Populus balsamifera.*   Pleistocene, Canada.

Hudson Bay territories.   It is a northern though not arctic species in its geographical range.   According to Lyell it occurs associated with the bones of Mastodon in New Jersey.   From the great durability of its wood, it is one of the trees most likely to be preserved in aqueous deposits.

7. *Potamogeton perfoliatus*, Linn.   Leaves and seeds in nodules at Green's Creek.   Inhabits streams of the Northern States and Canada, and according to Richardson extends to Great Slave Lake.

8. *Potamogeton pusillus.*   Quantities of fragments which I refer to this species occur in nodules at Green's Creek.   They may possibly belong to a variety of *P. hybridus* which, together with *P. natans*, now grows in

the river Ottawa, where it flows over the beds containing these fossils.

9. *Cariceæ* and *Gramineæ*. Fragments in nodules from Green's Creek appear to belong to plants of these groups, but I cannot venture to determine their species.

10. *Equisetum scirpoides*, Michx. Fragments in nodules, Green's Creek. This is a widely distributed species, occurring in the Northern States and Canada.

11. *Fontinalis*. In nodules at Green's Creek there occur, somewhat plentifully, branches of a moss apparently of the genus *Fontinalis*.

Fig. 79.—Frond of *Fucus.* Pleistocene, Canada.

12. *Algæ*. With the plants above mentioned, both at Green's Creek and at Montreal, there occur remains of seaweeds (Fig. 79). They seem to belong to the genera *Fucus* and *Ulva*, but I cannot determine the species. A thick stem in one of the nodules would seem to indicate a large *Laminaria*. With the above there are found at Green's Creek a number of fragments of leaves, stems, and fruits, which I have not been able to refer to their species, principally on account of their defective state of preservation.

None of the plants above mentioned is properly arctic in its distribution, and the assemblage may be characterised as a selection from the present Canadian flora of some of the more hardy species having the most northern range. Green's Creek is in the central part of Canada, near to the parallel of 46°, and an accidental selection

from its present flora, though it might contain the same
species found in the nodules, would certainly include with
these, or instead of some of them, more southern forms.
More especially the balsam poplar, though that tree occurs plentifully on the Ottawa, would not be so pre-
dominant.   But such an assemblage of drift-plants might
be furnished by any American stream flowing in the latitude of 50° to 55° north.   If a stream flowing to the
north, it might deposit these plants in still more northern
latitudes, as the McKenzie River does now.   If flowing
to the south, it might deposit them to the south of 50°.
In the case of the Ottawa, the plants could not have been
derived from a more southern locality, nor probably from
one very far to the north.   We may therefore safely assume that the refrigeration indicated by these plants
would place the region bordering the Ottawa in nearly the
same position with that of the south coast of Labrador
fronting on the Gulf of St. Lawrence at present.   The
absence of all the more arctic species occurring in Labrador should perhaps induce us to infer a somewhat
milder climate than this.

The moderate amount of refrigeration thus required
would in my opinion accord very well with the probable
conditions of climate deducible from the circumstances in
which the fossil plants in question occur.   At the time
when they were deposited the sea flowed up the Ottawa
valley to a height of 200 to 400 feet above its present
level, and the valley of the St. Lawrence was a wide arm
of the sea, open to the arctic current.   Under these conditions the immense quantities of drift-ice from the
northward, and the removal of the great heating surface
now presented by the low lands of Canada and New England, must have given for the Ottawa coast of that period
a summer temperature very similar to that at present experienced on the Labrador coast, and with this conclusion
the marine remains of the Leda clay, as well as the few

land molluscs whose shells have been found in the beds containing the plants, and which are species still occurring in Canada, perfectly coincide.

The climate of that portion of Canada above water at the time when these plants were imbedded may safely be assumed to have been colder in summer than at present, to an extent equal to about 5° of latitude, and this refrigeration may be assumed to correspond with the requirements of the actual geographical changes implied. In other words, if Canada was submerged until the Ottawa valley was converted into an estuary inhabited by species of *Leda*, and frequented by capelin, the diminution of the summer heat consequent on such depression would be precisely suitable to the plants occurring in these deposits, without assuming any other cause of change of climate.

I have arranged elsewhere the Post-pliocene deposits of the central part of Canada, as consisting of, in ascending order : (1) The boulder clay ; (2) a deep-water deposit, the Leda clay ; and (3) a shallow-water deposit, the Saxicava sand. But, although I have placed the boulder clay in the lowest position, it must be observed that I do not regard this as a continuous layer of equal age in all places. On the contrary, though locally, as at Montreal, under the Leda clay, it is in other places and at other levels contemporaneous with or newer than that deposit, which itself also locally contains boulders.

At Green's Creek the plant-bearing nodules occur in the lower part of the Leda clay, which contains a few boulders, and is apparently in places overlaid by large boulders, while no distinct boulder clay underlies it. The circumstances which accumulated the thick bed of boulder clay near Montreal were probably absent in the Ottawa valley. In any case we must regard the deposits of Green's Creek as coeval with the Leda clay of Montreal, and with the period of the greatest abundance of *Leda* *glacialis,* the most exclusively arctic shell of these deposits. In other words, I regard the plants above mentioned as probably belonging to the period of greatest refrigeration of which we have any evidence, of course not including that mythical period of universal incasement in ice, of which, as I have elsewhere endeavoured to show, in so far as Canada is concerned, there is no evidence whatever.*

The facts above stated in reference to Post-pliocene plants concur, with all the other evidence I have been able to obtain, in the conclusion that the refrigeration of Canada in the Post-pliocene period consisted of a diminution of the summer, heat, and was of no greater amount than that fairly attributable to the great depression of the land and the different distribution of the ice-bearing arctic current.

In connection with the plants above noticed, it is interesting to observe that at Green's Creek, at Pakenham Mills, at Montreal, and at Clarenceville on Lake Champlain, species of Canadian *Pulmonata* have been found in deposits of the same age with those containing the plants. The species which have been noticed belong to the genera *Lymnea* and *Planorbis.*

The Glacial age was, fortunately, not of very long duration, though its length has been much exaggerated by certain schools of geologists.† It passed away, and a returning cosmic spring gladdened the earth, and was ushered in by a time of great rainfall and consequent denudation and deposit, which has been styled the "Pluvial Period." The remains of the Pliocene forests then returned—with somewhat diminished numbers of species—

---

* Notes on Post-Pliocene of Canada, "Canadian Naturalist," 1872.

† This I have long maintained on grounds connected with Pleistocene fossils, amount of denudation and deposit, &c., and I am glad to see that Prestwich, the best English authority on such subjects, has recently announced similar conclusions, based on independent reasons.

from the south and again occupied the land, though they
have not been able, in their decimated condition, to restore the exuberance of the flora of the earlier Tertiary.
In point of fact, as we shall see in the next chapter, it is
the floras originating within the polar circle and coming
down from the north that are rich and copious. Those
that, after periods of cold or submergence, return from
the south, are comparatively poor. Hence the modern
flora is far inferior to that of the Middle Kainozoic. In
America, however, and in eastern Asia, for reasons already stated, the return was more abundant than in
Europe.

Simultaneously with the return of the old temperate
flora, the arctic plants that had overspread the land retreated to mountain-tops, now bared of ice and snow, and
back to the polar lands whence they came ; and so it happens that, on the White Mountains, the Alps, and the
Himalayas, we have insular patches of the same groups of
plants that exist around the pole.

These changes need not have required a very long
time, for the multiplication and migration of plants are
very rapid, especially when aided by the agency of migratory animals. Many parts of the land must, indeed, have
been stocked with plants from various sources, and by
agencies—as that of the sea—which might at first sight
seem adverse to their distribution. The British Islands,
for example, have no indigenous plants. Their flora
consists mainly of Germanic plants, which must have
migrated to Britain in that very late period of the Post-
glacial when the space now occupied by the North Sea
was mostly dry land. Other portions of it are Scandinavian plants, perhaps survivors of the Glacial age, or
carried by migratory birds ; and still another element
consists of Spanish plants, brought north by spring mi-
grants, and establishing themselves in warm and sheltered
spots, just as the arctic plants do on the bleak hill-tops.

The Bermudas, altogether recent islands, have one hundred and fifty species of native plants, all of which are West Indian and American, and must have been introduced by the sea-currents or by migratory birds.

And so the earth became fitted for the residence of modern man. Yet it is not so good or Edenic a world as it once was, or as it may yet become, were another revolution to restore a mild climate to the arctic regions, and to send down a new swarm of migratory species to renew the face of the earth and restore it to its pristine fertility of vegetable life.

Thus closes this long history of the succession of plants, reaching from the far back Laurentian to the present day. It has, no doubt, many breaks, and much remains to be discovered. Yet it may lead us to some positive conclusions regarding the laws of the introduction of plants.

One of these, and perhaps the most remarkable of all, is that certain principles were settled very far back, and have remained ever since. We have seen that in the earliest geological periods all that pertains to the structure, powers, and laws of the vegetable cell was already fixed and settled. When we consider how much this implies of mechanical structure and chemical and vital property, the profound significance of this statement becomes apparent. The relations in these respects between the living cell and the soil, the atmosphere and the sunshine, were apparently as perfect in the early Palæozoic as in any subsequent time. The same may be said of the structures of the leaf and of the stem. In such old forms as *Nematophyton* these were, it is true, peculiar and rudimentary, but in the Devonian and Carboniferous the structure of leaves and stems embodied all the parts and principles that we find at present. In regard to fructification there has been more progress, for, so far as we know, the highest and most complex forms of flowers, fruits, and seeds belong to the more recent periods, and simpler forms were at least dominant in the older times. Yet even in this respect the great leading laws and structures of bisexual reproduction were perfected in the early Palæozoic, and the improvements introduced in the gymnosperm and the angiosperm of later periods have consisted mainly in additions of accessory parts, and in modifications and refinements suited to the wants of the higher and more complex types.

CHAPTER VIII.

GENERAL LAWS OF ORIGIN AND MIGRATIONS OF PLANTS.
—RELATIONS OF RECENT AND FOSSIL FLORAS.

THE origination of the successive floras which have occupied the northern hemisphere in geological time, not, as one might at first sight suppose, in the sunny climes of the south, but under the arctic skies, is a fact long known or suspected. It is proved by the occurrence of fossil plants in Greenland, in Spitzbergen, and in Grinnell Land, under circumstances which show that these were their primal homes. The fact bristles with physical difficulties, yet is fertile of the most interesting theoretical deductions, to reach which we may well be content to wade through some intricate questions. Though not at all a new fact, its full significance seems only recently to have dawned on the minds of geologists, and within the last few years it has produced a number of memoirs and addresses to learned societies, besides many less formal notices.*

The earliest suggestion on the subject known to the writer is that of Prof. Asa Gray, in 1867, with reference to the probable northern source of the related floras of North America and eastern Asia. With the aid of the new facts disclosed by Heer and Lesquereux, Gray re-

---

* Saporta, "Ancienne Végétation Polaire"; Hooker, "Presidential Address to Royal Society," 1878; Thistleton Dyer, "Lecture on Plant Distribution"; Mr. Starkie Gardner, "Letters in 'Nature,'" 1878, &c. The basis of most of these brochures is to be found in Heer's "Flora Fossilis Arctica."

turned to the subject in 1872, and more fully developed this conclusion with reference to the Tertiary floras,* and he has recently still further discussed these questions in an able lecture on "Forest Geography and Archæology." † In this he puts the case so well and tersely that we may quote the following sentences as a text for what follows :

"I can only say, at large, that the same species (of Tertiary fossil plants) have been found all round the world ; that the richest and most extensive finds are in Greenland ; that they comprise most of the sorts which I have spoken of, as American trees which once lived in Europe—magnolias, sassafras, hickories, gum-trees, our identical southern cypress (for all we can see of difference), and especially *Sequoias*, not only the two which obviously answer to the two big-trees now peculiar to California, but several others ; that they equally comprise trees now peculiar to Japan and China, three kinds of gingko-trees, for instance, one of them not evidently distinguishable from the Japan species which alone survives ; that we have evidence, not merely of pines and maples, poplars, birches, lindens, and whatever else characterise the temperate zone forests of our era, but also of particular species of these, so like those of our own time and country that we may fairly reckon them as the ancestors of several of ours.   Long genealogies always deal more or less in conjecture ; but we appear to be within the limits of scientific inference when we announce that our existing temperate trees came from the north, and within the bounds of nigh probability when we claim not a few of them as the originals of present species.   Remains of the same plants have been found fossil in our temperate region as well as in Europe."

---

* Address to American Association.
† "American Journal of Science," xvi., 1878.

Between 1860 and 1870 the writer was engaged in working out all that could be learned of the Devonian plants of eastern America, the oldest known flora of any richness, and which consists almost exclusively of gigantic, and to us grotesque, representatives of the club-mosses, ferns, and mares'-tails, with some trees allied to the cycads and pines.   In this pursuit nearly all the more important localities were visited, and access was had to the large collections of Prof. Hall and Prof. Newberry, in New York and Ohio, and to those made in the remarkable plant-bearing beds of New Brunswick by Messrs. Matthew and Hartt.   In the progress of these researches, which developed an unexpectedly rich assemblage of species, the northern origin of this old flora seemed to be established by its earlier culmination in the northeast, in connection with the growth of the American land to the southward, which took place after the great Upper Silurian subsidence, by elevations beginning in the north while those portions of the continent to the southwest still remained under the sea.   The same result was indicated by the persistence in the Carboniferous of the south and west of old Erian forms, like *Megalopteris*.

When, in 1870, the labours of those ten years were brought before the Royal Society of London, in the Bakerian lecture of that year, and in a memoir illustrating no less than one hundred and twenty-five species of plants older than the great Carboniferous system, these deductions were stated in connection with the conclusions of Hall, Logan, and Dana, as to the distribution of sediment along the northeast side of the American continent, and the anticipation was hazarded that the oldest Palæozoic floras would be discovered to the north of Newfoundland.   Mention was also made of the apparent earlier and more copious birth of the Devonian flora in America than in Europe, a fact which is itself connected with the greater northward extension of this continent.

The memoir containing these results was not published
by the Royal Society, but its publication was secured in a
less complete form in the reports of the "Geological Survey of Canada." The part of the memoir relating to Canadian fossil plants, with a portion of the theoretical deductions, was published in a report issued in 1871.* In this
report the following language was used :

"In eastern America, from the Carboniferous period
onward, the centre of plant distribution has been the Appalachian chain. From this the plants and sediments
extended westward in times of elevation, and to this they
receded in times of depression. But this centre was non-
existent before the Devonian period, and the centre for
this must have been to the northeast, whence the great
mass of older Appalachian sediment was derived. In the
Carboniferous period there was also an eastward distribution from the Appalachians, and links of connection in
the Atlantic bed between the floras of Europe and America. In the Devonian such connection can have been only
far to the northeast. It is therefore in Newfoundland,
Labrador, and Greenland that we are to look for the
oldest American flora, and in like manner on the border
of the old Scandinavian nucleus for that of Europe.

"Again, it must have been the wide extension of the
sea of the corniferous limestone that gave the last blow
to the remaining flora of the Lower Devonian ; and the
re-elevation in the middle of that epoch brought in the
Appalachian ridges as a new centre, and established a
connection with Europe which introduced the Upper
Devonian and Carboniferous floras. Lastly, from the
comparative richness of the later Erian † flora in eastern
America, especially in the St. John beds, it might be a

---

* "Fossil Plants of the Devonian and Upper Silurian Formations of
Canada," pp. 92, twenty plates, Montreal, 1871.

† See pages 107 and 108.

fair inference that the northeastern end of the Appalachian ridge was the original birthplace or centre of creation of what we may call the later Palæozoic flora, or of a large part of that flora."

When my paper was written I had not seen the account published by the able Swiss palæobotanist Heer, of the remarkable Devonian flora of Bear Island, near Spitzbergen.* From want of acquaintance with the older floras of America and western Europe, Heer fell into the unfortunate error of regarding the whole of Bear Island plants as Lower Carboniferous, a mistake which his great authority has tended to perpetuate, and which has even led to the still graver error of some European geologists, who do not hesitate to regard as Carboniferous the fossil plants of the American deposits from the Hamilton to the Chemung groups inclusive, though these belong to formations underlying the oldest Carboniferous, and characterised by animal remains of unquestioned Devonian age. In 1872 I addressed a note to the Geological Society of London on the subject of the so-called "Ursa stage" of Heer, showing that, though it contained some forms not known at so early a date in temperate Europe, it was clearly, in part at least, Devonian when tested by North American standards; but that in this high latitude, in which, for reasons stated in the report above referred to, I believed the Devonian plants to have originated, there might be an intermixture of the two floras. But such a mixed group should in that latitude be referred to a lower horizon than if found in temperate regions. Dr. Nathorst, as already stated, has recently obtained new facts which go to show that plants of two distinct horizons may have been intermixed in the collections submitted to Heer.

---

* "Transactions of the Swedish Academy," 1871; "Journal of the London Geological Society," vol. xxviii.

Between 1870 and 1873 my attention was turned to the two subfloras intermediate between those of the Devonian and the coal-formation, the floras of the Lower Carboniferous (Subcarboniferous of some American geologists) and the Millstone Grit, and in a report upon these * similar deductions were expressed. It was stated that in Newfoundland the coal-beds seem to belong to the Millstone Grit series, and as we proceed southward they belong to progressively newer portions of the Carboniferous system. The same fact is observed in the coal-beds of Scotland, as compared with those of England, and it indicates that the coal-formation flora, like that of the Devonian, spread itself from the north, and this accords with the somewhat extensive occurrence of Lower Carboniferous rocks and fossils in the Parry Islands and elsewhere in the arctic regions.

Passing over the comparatively poor flora of the earlier Mesozoic, consisting largely of cycads, pines, and ferns, and as yet little known in the arctic, and which may have originated in the south, though represented, according to Heer, by the supposed Jurassic flora of Siberia, we find, especially at Komé and Atané in Greenland, an interesting occurrence of those earliest precursors of the truly modern forms of plants which appear in the Cretaceous, the period of the English chalk and of the New Jersey greensands. There are two plant-groups of this age in Greenland; one, that of Komé, consists almost entirely of ferns, cycads, and pines, and is of decidedly Mesozoic aspect. This is called Lower Cretaceous. The other, that of Atané, holds remains of many modern temperate genera, as *Populus*, *Myrica*, *Ficus*, *Sassafras*, and *Magnolia*. This is regarded as Upper Cretaceous. Resting upon these Upper Cretaceous beds, without the inter-

---

* "Fossil Plants of Lower Carboniferous and Millstone Grit Formations of Canada," pp. 47, ten plates, Montreal, 1873.

vention of any other formation,* are beds rich in plants
of much more modern appearance, and referred by Heer
to the Miocene period, a reference, as we have seen, not
warranted by comparison with the Tertiary plants of Europe or of America.   Still farther north this so-called
Miocene assemblage of plants appears in Spitzbergen and
Grinnell Land ; but there, owing to the predominance of
trees allied to the spruces, it has a decidedly more boreal
character than in Greenland, as might be anticipated from
its nearer approach to the pole.†

If now we turn to the Cretaceous and Tertiary floras
of western America, as described by Lesquereux, Newberry, and others, we find in the lowest Cretaceous rocks
there known—those of the Dakota group—which may be
in the lower part of the Middle Cretaceous, a series of
plants‡ essentially similar to those of the so-called Upper
Cretaceous of Greenland.   They occur in beds indicating
land and fresh-water conditions as prevalent at the time
over great areas of the interior of America.   But overlying this plant-bearing formation we have an oceanic
limestone (the Niobrara), corresponding in many respects
to the European chalk, and extending far north into the
British territory,# indicating that the land of the Lower
Cretaceous was replaced by a vast Mediterranean Sea,
filled with warm water from the equatorial currents, and
not invaded by cold waters from the north.   This is succeeded by thick Upper Cretaceous deposits of clay and
sandstone, with marine remains, though very sparsely

---

* Nordenskiöld, "Expedition to Greenland," "Geological Magazine,"
1872.

† Yet even here the bald cypress (*Taxodium distichum*), or a tree
nearly allied to it, is found, though this species is now limited to the
Southern States.   Fielden and De Rance, "Journal of the Geological Society," 1878.

‡ Lesquereux, "Report on Cretaceous Flora."

G. M. Dawson, "Report on Forty-ninth Parallel."

distributed ; and these show that further subsidence or denudation in the north had opened a way for the arctic currents, killing out the warm-water animals of the Niobrara group, and filling up the Mediterranean of that period. Of the flora of these Upper Cretaceous periods, which must have been very long, we know something in the interior regions, from the discovery of a somewhat rich flora in the Dunvegan beds of the Peace River district, on the northern shore of the great Cretaceous Mediterranean ;* and on the coast of British Columbia we have the remarkable Cretaceous coal-field of Vancouver Island, which holds the remains of plants of modern genera, and, indeed, of almost as modern aspect as those of the so-called Miocene of Greenland. They indicate, however, a warmer climate as then prevalent on the Pacific coast, and in this respect correspond with a peculiar transition flora, intermediate between the Cretaceous and Eocene or earliest Tertiary of the interior regions, and which is described by Lesquereux as the Lower Lignitic.

Immediately above these Upper Cretaceous beds we have the great Lignite Tertiary of the West—the Laramie group of recent American reports—abounding in fossil plants, at one time regarded as Miocene, but now known to be Lower Eocene, though farther south extending upward toward the Miocene age.† These beds, with their characteristic plants, have been traced into the British territory north of the forty-ninth parallel, and it has been shown that their fossils are identical with those of the

---

* "Reports of Dr. G. M. Dawson, Geological Survey of Canada." Also, "Transactions of the Royal Society of Canada," vol. i.

† Lesquereux's "Tertiary Flora"; "White on the Laramie Group"; Stevenson, "Geological Relations of Lignitic Groups," American Philosophical Society, June, 1875 ; Dawson, "Transactions of the Royal Society of Canada," vol. iv.; Ward, "Bulletin of United States Geological Survey."

McKenzie River valley, described by Heer as Miocene, and probably also with those of Alaska, referred to the same age.*   Now this truly Eocene flora of the temperate and northern parts of America has so many species in common with that called Miocene in Greenland that its identity can scarcely be doubted.   These facts have led to scepticism as to the Miocene age of the upper plant-bearing beds of Greenland, and more especially Mr. J. Starkie Gardner has ably argued, from comparison with the Eocene flora of England and other considerations, that they are really of that earlier date.†

In looking at this question, we may fairly assume that no climate, however equable, could permit the vegetation of the neighbourhood of Disco in Greenland to be exactly identical with that of Colorado and Missouri, at a time when little difference of level existed in the two regions.   Either the southern flora migrated north in consequence of a greater amelioration of climate, or the northern flora moved southward as the climate became colder.   The same argument, as Gardner has ably shown, applies to the similarity of the Tertiary plants of temperate Europe to those of Greenland.   If Greenland required a temperature of about 50°, as Heer calculates, to maintain its Eocene flora, the temperature of England and that of the Southwestern States must have been higher, though probably more equable, than at present.

We cannot certainly affirm anything respecting the migrations of these floras, but there are some probabilities which deserve attention.   The ferns and cycads of the so-called Lower Cretaceous of Greenland are nothing but a continuation of the previous Jurassic flora.   Now this was established at an equally early date in the Queen

---

* G. M. Dawson, " Report on the Geology of the Forty-ninth Parallel," where full details on these points may be found. " Transactions of the Royal Society of Canada," vol. iv.

† " Nature," December 12, 1878.

Charlotte Islands,* and still earlier in Virginia.† The presumption is, therefore, that it came from the south. It has, indeed, the facies of a southern hemisphere and insular flora, and probably spread itself northward as far as Greenland, at a time when our northern continents were groups of islands, and when the ocean currents were carrying warm water far toward the arctic regions. The flora which succeeds this in the sections at Atané has no, special affinities with the southern hemisphere, and is of a more temperate and continental character.‡ It is not necessarily Upper Cretaceous, since it is similar to that of the Dakota group farther south, and this is at least Middle Cretaceous. This flora must have originated either somewhere in temperate America or within the Arctic circle, and it must have replaced the older one by virtue of increasing coolness and continental character of climate. It must, therefore, have been connected with that elevation of the land which took place at the beginning of the Cretaceous. During this elevation it spread over all western America at one time or another, and, as the land again subsided under the sea of the Niobrara chalk, it assumed an aspect more suited to a warm climate, but still held its place on such islands as remained above water along the Pacific coast and in the north, and it continued to exist on these islands till the colder seas

---

* "Reports of the Geological Survey of Canada."

† Fontaine has well described the Mesozoic flora of Virginia, "American Journal of Science," January, 1879, and "Report on Early Mesozoic Floras."

‡ In the "Proceedings of the Royal Society of Tasmania," 1887, Mr. R. M. Johnston, F. L. S., states that in the Miocene beds of Tasmania trees of European genera abound. The Mesozoic flora of that island is of the usual conifero-cycadean type. Ettingshausen makes a similar statement in the "Geological Magazine" respecting the Tertiary flora of Australia and New Zealand, stating that, like the Tertiary floras of Europe, they have a mixed character, being partly of types now belonging to the northern hemisphere.

of the Upper Cretaceous had again given place to the warm plains and land-locked brackish seas or fresh-water lakes of the Laramie period (Eocene).   Thus the true Upper Cretaceous marks a cool period intervening between the so-called Upper Cretaceous (really Middle Cretaceous) and the so-called Miocene (really Lower Eocene) floras of Greenland.

This latter established itself in Greenland, and probably all around the Arctic circle, in the warm period of the earliest Eocene, and, as the climate of the northern hemisphere became gradually reduced from that time till the end of the Pliocene, it marched on over both continents to the southward, chased behind by the modern arctic flora, and eventually by the frost and snow of the Glacial age.   This history may admit of correction in details; but, so far as present knowledge extends, it is in the main not far from the truth.

Perhaps the first great question which it raises is that as to the causes of the alternations of warm and cold climates in the north, apparently demanded by the vicissitudes of the vegetable kingdom.   Here we may set aside the idea that in former times plants were suited to endure greater cold than at present.   It is true that some of the fossil Greenland plants are of unknown genera, and many are species new to us; but we are on the whole safe in affirming that they must have required conditions similar to those necessary to their modern representatives, except within such limits as we now find to hold in similar cases among existing plants.   Still we know that at the present time many species found in the equable climate of England will not live in Canada, though species to all appearance similar in structure are native here.   There is also some reason to suppose that species when new may have greater hardiness and adaptability than when in old age and verging toward extinction.   In any case these facts can account for but a small part of the phenomena, which require to be explained by physical changes affecting the earth as a whole, or at least the northern hemisphere. Many theoretical views have been suggested on this subject, and perhaps the most practical way of disposing of these will be first to set aside a number which are either precluded by the known facts, incapable of producing the effects, or altogether uncertain as to their possible occurrence.

1. In this class we may place the theory that the poles of the earth have changed their position. Independently of astronomical objections, there is good geological evidence that the poles of the earth must have been nearly in their present places from the dawn of life until now. From the Laurentian upward, those organic limestones which mark the areas where warm and shallow equatorial water was spreading over submerged continents are so disposed as to prove the permanence of the poles. In like manner all the great foldings of the crust of the earth have followed lines which are parts of great circles tangent to the existing polar circles. So, also, from the Cambrian age the great drift of sediment from the north has followed the line of the existing Arctic currents from the northeast to the southwest, throwing itself, for example, along the line of the Appalachian uplifts in eastern America, and against the ridge of the Cordilleras in the west.

2. Some of the above considerations, along with astronomical evidence, prevent us from assuming any considerable change in the obliquity of the axis of the earth during geological time.

3. That the earth and the sun have diminished in heat during geological time seems probable ; but physical and geological facts alike render it certain that this influence could have produced no appreciable effect, even in the times of the earliest floras, and certainly not in the case of Tertiary vegetation.

4. It has been supposed that the earth may have at different times traversed more or less heated zones of space, giving alternations of warm and cold temperature. No such differences in space are, however, known, nor does there seem any good ground for imagining their existence.

5. The heat of the sun is known to be variable, and the eleven years' period of sun-spots has recently attracted much attention as producing appreciable effects on the seasons.   There may possibly be longer cycles of solar energy, or the sun may be liable, like some variable stars, to paroxysms of increased energy.   Such changes are possible, and may fairly be taken into the account, provided that we fail to find known causes sufficient to account for the phenomena.

Of well-known causes there seem to be but three. These are : First, that urged by Lyell—viz., the varying distribution of land and water along with that of marine currents ; secondly, the varying eccentricity of the earth's orbit, along with the precession of the equinoxes, and the effects of this on oceanic circulation, as illustrated by Croll ; thirdly, the different conditions of the earth's atmosphere with reference to radiation, as argued by Tyndall and Hunt.   As these causes are all founded on known facts, and not exclusive of each other, we may consider them together.   I shall take the Lyellian theory first, regarding it as the most important, and the best supported by geological facts.

We know that the present distribution of land and water greatly influences climate, more especially by affecting that of the ocean currents and of the winds, and by the different action of land as compared with water in the reception and radiation of heat.   The present distribution of land gives a large predominance to the arctic and sub-arctic regions, as compared with the equatorial and with the antarctic ; and we might readily imagine

other distributions that would give very different results.
But this is not an imaginary case. We know that, while
the forms and positions of the great continents have been
fixed from a very early date, they have experienced many
great submergences and re-elevations, and that these have
occurred in somewhat regular sequence, as evidenced by
the cyclical alternations of organic limestones and earthy
sediments in successive geological formations.

An example bearing on our present subject may serve
to illustrate this. In the latter part of the Upper Silurian period (the Lower Helderberg age), vast areas of the
American continent * were covered with an ocean in
which were deposited organic limestones whose fossils
show that this great interior sea was pervaded by equatorial waters bringing food and warmth, while the in-
cipient ranges of the Appalachians on the east, and the
Cordilleras on the west, and the Laurentian axis on the
north, fenced off from it the colder arctic waters. How
different must the climate of America and of the region
north of it have been in these circumstances from that
which prevails at present, or from that which prevailed
in certain other periods, when it was open to the incur-
sions of the arctic ice-laden currents, bearing loads of fine
sediment ! † It was in these circumstances, and in the
similar circumstances in which the great Corniferous
limestone of the Devonian was deposited—a limestone
showing in its rich coral fauna even warmer waters than
those of the Lower Helderberg—that the Devonian flora

---

* See a memoir and map by Prof. Hall, "Reports of the Regents of
New York," 1874–'75.

† It seems certain that the faunæ of the old limestones, like the Tren-
ton, Niagara, Lower Helderberg, and Corniferous, belong to warm and
sheltered sea areas, and that those rich in graptolites and trilobites, enclosed in muddy sediments, belong to the colder arctic waters. Such
arctic faunæ are those of the Quebec group and of the Utica shale, and
to some extent that of the Hamilton group.

took its origin in the north and advanced southward over new lands in process of emergence from the sea. The somewhat similar condition evidenced by the Lower Carboniferous limestone preceded the advent of the great and rich flora of the coal-formation.

Lyell's theory on this subject has, I think, in some recent publications, been somewhat misapprehended. It is true that he stated hypothetically two contrasted conditions of distribution, in one of which all the land was equatorial, in another all polar; but he did not suppose that these conditions had actually occurred; and even in his earlier editions, before the recent discoveries and discussions as to ocean currents, he was always careful to attach due value to these in connection with subsidences and elevations.* In his later editions he introduced more full references to current action, and also stated Croll's theory, but still maintained the validity of his original conclusions.

The sufficiency of this Lyellian theory to account for the facts, in so far as plants are concerned, may, I think, be inferred from the course of the isothermal lines at present. The south end of Greenland is on the latitude of Christiania in Norway on the one hand, and of Fort Liard in the Peace River region on the other; and while Greenland is clad in ice and snow, wheat and other grains, and the ordinary trees of temperate climates, grow at the latter places.† It is evident, therefore, that only exceptionally unfavourable circumstances prevent the Greenland area from still possessing a temperate flora, and these unfavourable circumstances possibly tell even on the localities with which we have compared it. Further, the mouth of the McKenzie River is in the same latitude with

---

* See " Principles of Geology," edition of 1840, chapter vii.

† See " Macoun's Report," " Geological Survey of Canada," and Richardson's " Boat Voyage."

Disco, near which are some of the most celebrated localities of fossil Cretaceous and Tertiary plants. Yet the mouth of the McKenzie River enjoys a much more favourable climate and has a much more abundant flora than Disco. If north Greenland were submerged, and low land reaching to the south terminated at Disco, and if from any cause either the cold currents of Baffin's Bay were arrested, or additional warm water thrown into the North Atlantic by the Gulf Stream, there is nothing to prevent a mean temperature of 45° Fahr. from prevailing at Disco; and the estimate ordinarily formed of the requirements of its extinct floras is 50°,[*] which is probably above rather than below the actual temperature required.

Since, then, geological facts assure us of mutations of the continents much greater than those apparently required to account for the changes of climate implied in the existence of the ancient arctic floras, it does not seem absolutely necessary to invoke any others.[†] If, however, there are other true causes which might either aid or counteract those above referred to, it may be well to consider them.

Mr. Croll has, in his valuable work "Climate and Time," and in various memoirs, brought forward an ingenious astronomical theory to account for changes of climate. This theory, as stated by himself in a recent paper,[‡] is that when the eccentricity of the earth's orbit is at a high value, and the northern winter solstice is in perihelion, agencies are brought into operation which make the southeast trade-winds stronger than the northeast, and compel them to blow over upon the northern

---

[*] Heer. See, also, papers by Prof. Haughton and by Gardner in "Nature" for 1878.

[†] Sir William Thomson, "Transactions of the Geological Society of Glasgow," February 22, 1878.

[‡] "Cataclysmic Theories of Geological Climate," "Geological Magazine," May, 1878.

hemisphere as far as the Tropic of Cancer.   The result is that all the great equatorial currents of the ocean are impelled into the northern hemisphere, which thus, in consequence of the immense accumulation of warm water, has its temperature raised, so that ice and snow must to a great extent disappear from the arctic regions.   In the prevalence of the converse conditions, the arctic zone becomes clad in ice, and the southern has its temperature raised.

At the same time, according to Croll's calculations, the accumulation of ice on either pole would tend, by shifting the earth's centre of gravity, to raise the level of the ocean and submerge the land on the colder hemisphere. Thus a submergence of land would coincide with a cold condition, and emergence with increasing warmth.   Facts already referred to, however, show that this has not always been the case, but that in many cases submergence was accompanied with the influx of warm equatorial waters and a raised temperature, this apparently depending on the question of local distribution of land and water ; and this in its turn being regulated not always by mere shifting of the centre of gravity, but by foldings occasioned by contraction, by equatorial subsidences resulting from the retardation of the earth's rotation, and by the excess of material abstracted by ice and frost from the arctic regions, and drifted southward along the lines of arctic currents.   This drifting must in all geological times have greatly exceeded, as it certainly does at present, the denudation caused by atmospheric action at the equator, and must have tended to increase the disposition to equatorial collapse occasioned by retardation of rotation.*

While such considerations as those above referred to

---

* Croll, in "Climate and Time," and in a note read before the British Association in 1876, takes an opposite view; but this is clearly contrary to the facts of sedimentation, which show a steady movement of *débris* toward the south and southwest.

tend to reduce the practical importance of Mr. Croll's theory, on the other hand they tend to remove one of the greatest objections against it—namely, that founded on the necessity of supposing that glacial periods recur with astronomical regularity in geological time. They cannot do so if dependent on other causes inherent in the earth itself, and producing important movements of its crust.

The third great cause of warmer climates in the past is the larger proportion of carbon dioxide, or carbonic-acid gas, in the atmosphere in early geological times, as proved by the immense amount of carbon now sealed up in limestone and coal, and which must at one time have been in the air. It has been shown that a very small additional quantity of this substance would so obstruct radiation of heat from the earth as to act almost like a glass roof. If, however, the quantity of carbonic acid, great at first, was slowly and regularly removed, even if, as suggested by Hunt, small additional supplies were gradually added from space, this cause could have affected only the very oldest floras. But it is known that some comets and meteorites contain carbonaceous matter, and this allows us to suppose that accessions of carbon may have been communicated at irregular intervals. If so, there may have been cycles of greater and less abundance of this substance, and an atmosphere rich in carbon dioxide might at one and the same time afford warmth and abundance of food to plants.

It thus appears that the causes of ancient vicissitudes of climate are somewhat complex, and when two or more of them happened to coincide very extreme changes might result, having most important bearings on the distribution of plants.

This may help us to deal with the peculiarities of the great Glacial age, which may have been rendered exceptionally severe by the combination of several of the causes of refrigeration. We must not suppose, however, that

the views of those extreme glacialists who suppose continental ice-caps reaching half way to the equator are borne
out by facts.   In truth, the ice accumulating round the
pole must have been surrounded by water, and there must
have been tree-clad islands in the midst of the icy seas,
even in the time of greatest refrigeration.   This is proved
by the fact that, in the Leda clay of eastern Canada,
which belongs to the time of greatest submergence, and
whose fossil shells show sea-water almost at the freezing-
point, there are leaves of poplars and other plants which
must have been drifted from neighbouring shores.   Similar remains occur in clays of like origin in the basin of
the great lakes and in the West.   These have been called
" interglacial," but there is no evidence to prove that they
are not truly glacial.   Thus, while we need not suppose
that plants existed within the Arctic circle in the Glacial
age, we have evidence that those of the cold temperate
and sub-arctic zones continued to exist pretty far north.
At the same time the warm temperate flora would be
driven to the south, except where sustained in insular
spots warmed by the equatorial currents.   It would return
northward on the re-elevation of the land and the re-
newal of warmth.

If, however, our modern flora is thus one that has returned from the south, this would account for its poverty
in species as compared with those of the early Tertiary.
Groups of plants descending from the north have been
rich and varied.   Returning from the south they are like
the shattered remains of a beaten army.   This, at least,
has been the case with such retreating floras as those of
the Lower Carboniferous, the Permian, and the Jurassic,
and possibly that of the Lower Eocene of Europe.

The question of the supply of light to an arctic flora
is much less difficult than some have imagined.   The
long summer day is in this respect a good substitute for
a longer season of growth, while a copious covering of

winter snow not only protects evergreen plants from those sudden alternations of temperature which are more destructive than intense frost, and prevents the frost from penetrating to their roots, but, by the ammonia which it absorbs, preserves their greenness. According to Dr. Brown, the Danish ladies of Disco long ago solved this problem.* He informs us that they cultivate in their houses most of our garden flowers—as roses, fuchsias, and geraniums—showing that it is merely warmth and not light that is required to enable a subtropical flora to thrive in Greenland. Even in Canada, which has a flora richer in some respects than that of temperate Europe, growth is effectually arrested by cold for nearly six months, and though there is ample sunlight there is no vegetation. It is, indeed, not impossible that in the plans of the Creator the continuous summer sun of the arctic régions may have been made the means for the introduction, or at least for the rapid growth and multiplication, of new and more varied types of plants.

Much, of course, remains to be known of the history of the old floras, whose fortunes I have endeavoured to sketch, and which seem to have been driven like shuttlecocks from north to south, and from south to north, especially on the American continent, whose meridional extension seems to have given a field specially suited for such operations.

This great stretch of the western continent, from north to south, is also connected with the interesting fact that, when new floras are entering from the arctic régions, they appear earlier in America than in Europe, and that in times when old floras are retreating from the south old genera and species linger longer in America. Thus, in the Devonian and Cretaceous new forms of those periods appear in America long before they are recognized

---

* "Florula Discoana," Botanical Society of Edinburgh, 1868.

in Europe, and in the modern epoch forms that would be
regarded in Europe as Miocene still exist.   Much confusion in reasoning as to the geological ages of the fossil floras
has arisen from want of attention to this circumstance.

What we have learned respecting this wonderful history has served strangely to change some of our precon-
ceived ideas.   We must now be prepared to admit that
an Eden can be planted even in Spitzbergen, that there
are possibilities in this old earth of ours which its present
condition does not reveal to us ; that the present state of
the world is by no means the best possible in relation to
climate and vegetation ; that there have been and might
be again conditions which could convert the ice-clad arctic regions into blooming paradises, and which at the
same time would moderate the fervent heat of the tropics.
We are accustomed to say that nothing is impossible with
God ; but how little have we known of the gigantic possibilities which lie hidden under some of the most common of his natural laws !

These facts have naturally been made the occasion of
speculations as to the spontaneous development of plants
by processes of varietal derivation.   It would, from this
point of view, be a nice question to calculate how many
revolutions of climate would suffice to evolve the first land-
plant ; what are the chances that such plant would be so
dealt with by physical changes as to be preserved and
nursed into a meagre flora like that of the Upper Silurian
or the Jurassic ; how many transportations to Greenland
would suffice to promote such meagre flora into the rich
and abundant forests of the Upper Cretaceous, and to
people the earth with the exuberant vegetation of the
early Tertiary.   Such problems we may never be able to
solve.   Probably they admit of no solution, unless we invoke the action of an Almighty mind, operating through
long ages, and correlating with boundless power and wisdom all the energies inherent in inorganic and organic

s

nature. Even then we shall perhaps be able to comprehend only the means by which, after specific types have been created, they may, by the culture of their Maker, be "sported" into new varieties or subspecies, and thus fitted to exist under different conditions or to occupy higher places in the economy of nature.

Before venturing on such extreme speculations as some now current on questions of this kind, we would require to know the successive extinct floras as perfectly as those of the modern world, and to be able to ascertain to what extent each species can change either spontaneously or under the influence of struggle for existence or expansion under favourable conditions, and under arctic semi-annual days and nights, or the shorter days of the tropics. Such knowledge, if ever acquired, it may take ages of investigation to accumulate.

As to the origin and mode of introduction of successive floras, I am, for the reasons above stated, not disposed to dogmatise, or to adopt as final any existing theory of the development of the vegetable kingdom. Still, some laws regulating the progress of vegetable life may be recognised, and I propose to state these in connection with the Palæozoic floras, to which my own studies have chiefly related.

Fossil plants are almost proverbially uncertain with reference to their accurate determination, and have been regarded as of comparatively little utility in the decision of general questions of palæontology. This results principally from the fragmentary condition in which they have been studied, and from the fact that fragments of animal structures are more definite and instructive than corresponding portions of plants.

It is to be observed, however, that our knowledge of fossil plants becomes accurate in proportion to the extent to which we can carry the study of specimens in the beds in which they are preserved, so as to examine more perfect examples than those usually to be found in museums. When structures are taken into the account, as well as external forms, we can also depend more confidently on our results. Further, the abundance of specimens to be obtained in particular beds often goes far to make up for their individual imperfection. The writer of these pages has been enabled to avail himself very fully of these advantages ; and on this account, if on no other, feels entitled to speak with some authority on theoretical questions.

It is an additional encouragement to pursue the subject, that, when we can obtain definite information as to the successive floras of any region, we thereby learn much as to climate and vicissitudes in regard to the extent of land and water ; and that, with reference to such points, the evidence of fossil plants, when properly studied, is, from the close relation of plants to those stations and climates, even more valuable than that of animal fossils.

It is necessary, however, that in pursuing such inquiries we should have some definite views as to the nature and permanence of specific forms, whether with reference to a single geological period or to successive periods ; and I may be excused for stating here some general principles, which I think important for our guidance.

1. Botanists proceed on the assumption, vindicated by experience, that, within the period of human observation, species have not materially varied or passed into each other. We may make, for practical purposes, the same assumption with regard to any given geological period, and may hold that for each such period there are specific types which, for the time at least, are invariable.

2. When we inquire what constitutes a good species for any given period, we have reason to believe that many names in our lists represent merely varietal forms or erroneous determinations. This is the case even in the modern flora ; and in fossil floras, through the poverty of specimens, their fragmentary condition, and various states

of preservation, it is still more likely to occur. Every revision of any group of fossils detects numerous synonyms, and of these many are incapable of detection without the comparison of large suites of specimens.

3. We may select from the flora of any geological period certain forms, which I shall call *specific types*, which may for such period be regarded as unchanging. Having settled such types, we may compare them with similar forms in other periods, and such comparisons will not be vitiated by the uncertainty which arises from the comparison of so-called species which may, in many cases, be mere varietal forms, as distinguished from specific types. Our types may be founded on mere fragments, provided that these are of such a nature as to prove that they belong to distinct forms which cannot pass into each other, at least within the limits of one geological period.

4. When we compare the specific types of one period with those of another immediately precedent or subsequent, we shall find that some continue unchanged through long intervals of geological time, that others are represented by allied forms regarded either as varietal or specific, and as derived or otherwise, according to the view which we may entertain as to the permanence of species. On the other hand, we also find new types not rationally deducible on any theory of derivation from those known in other periods. Further, in comparing the types of a poor period with those of one rich in species, we may account for the appearance of new types in the latter by the deficiency of information as to the former ; where many new types appear in the poorer period this conclusion seems less probable. For example, new types appearing in poor formations, like the Lower Erian and Lower Carboniferous, have greater significance than if they appeared in the Middle Erian or in the Coal Measures.

5. When specific types disappear without any known successors, under circumstances in which it seems un-

likely that we should have failed to discover their continuance, we may fairly assume that they have become
extinct, at least locally; and where the field of observation is very extensive, as in the great coal-fields of Europe
and America, we may esteem such extinction as practically general, at least for the northern hemisphere.
When many specific types become extinct together, or in
close succession, we may suppose that such extinction
resulted from physical changes; but where single types
disappear, under circumstances in which others of similar
habit continue, we may not unreasonably conjecture that,
as Pictet has argued in the case of animals, such types
may have been in their own nature limited in duration,
and may have died out without any external cause.

6. With regard to the *introduction* of specific types
we have not as yet a sufficient amount of information.
Even if we freely admit that ordinary specific forms, as
well as mere varieties, may result from derivation, this by
no means excludes the idea of primitive specific types
originating in some other way. Just as the chemist, after
analysing all compounds and ascertaining all allotropic
forms, arrives at length at certain elements not mutually
transmutable or derivable, so the botanist and zoologist
must expect sooner or later to arrive at elementary
specific types, which, if to be accounted for at all, must
be explained on some principle distinct from that of
derivation. The position of many modern biologists, in
presence of this question, may be logically the same with
that of the ancient alchemists with reference to the
chemical elements, though the fallacy in the case of fossils may be of more difficult detection. Our business at
present, in the prosecution of palæobotany, is to discover,
if possible, what are elementary or original types, and, having found these, to enquire as to the law of their creation.

7. In prosecuting such questions geographical relations must be carefully considered. When the floras of

two successive periods have existed in the same region, and under circumstances that render it probable that plants have continued to grow on the same or adjoining areas throughout these periods, the comparison becomes direct, and this is the case with the Erian and Carboniferous floras in northeastern America. But, when the areas of the two formations are widely separated in space as well as in time, any resemblances of facies that we may observe may have no connection whatever with an unbroken continuity of specific types.

I desire, however, under this head, to affirm my conviction that, with reference to the Erian and Carboniferous floras of North America and of Europe, the doctrine of "homotaxis," as distinct from actual contemporaneity, has no place. The succession of formations in the Palæozoic period evidences a similar series of physical phenomena on the grandest scale throughout the northern hemisphere. The succession of marine animals implies the continuity of the sea-bottoms on which they lived. The headquarters of the Erian flora in America and Europe must have been in connected or adjoining areas in the North Atlantic. The similarity of the Carboniferous flora on the two sides of the Atlantic, and the great number of identical species, proves a still closer connection in that period. These coincidences are too extensive and too frequently repeated to be the result of any accident of similar sequence at different times, and this more especially as they extend to the more minute differences in the features of each period, as, for instance, the floras of the Lower and Upper Devonian, and of the Lower, Middle, and Upper Carboniferous.

8. Another geographical question is that which relates to centres of dispersion. In times of slow subsidence of extensive areas, the plants inhabiting such areas must be narrowed in their range and often separated from one another in detached spots, while, at the same time, impor-

tant climatal changes must also occur. On the re-emergence of the land such of these species as remained would again extend themselves over their former areas of distribution, in so far as the new climatal and other conditions would permit. We would naturally suppose that the first of the above processes would tend to the elimination of varieties, the second, to their increase ; but, on the other hand, the breaking up of a continental flora into that of distinct islets, and the crowding together of many forms, might be a process fertile in the production of some varieties if fatal to others.

Further, it is possible that these changes of subsidence may have some connection with the introduction, as well as with the extinction, even of specific types. It is certain, at least, in the case of land-plants, that such types come in most plentifully immediately after elevation, though they are most abundantly preserved in periods of slow subsidence. I do not mean, however, that this connection is one of cause and effect ; there are, indeed, indications that it is not so. One of these is, that in some cases the enlargement of the area of the land seems to be as injurious to terrestrial species as its diminution.

9. Another point on which I have already insisted, and which has been found to apply to the Tertiary as well as to the Palæozoic floras, is the appearance of new types within the arctic and boreal areas, and their migration southward. Periods in which the existence of northern land coincided with a general warm temperature of the northern hemisphere seem to have been those most favourable to the introduction of new forms of land-plants. Hence, there has been throughout geological time a general movement of new floras from the Palæarctic and Nearctic regions to the southward.

Applying the above considerations to the Erian and Carboniferous floras of North America, we obtain some data which may guide us in arriving at general conclusions. The Erian flora is comparatively poor, and its types are in the main similar to those of the Carboniferous. Of these types a few only reappear in the middle coal-formation under identical forms; a great number appear under allied forms; some altogether disappear. The Erian flora of New Brunswick and Maine occurs side by side with the Carboniferous of the same region; so does the Erian of New York and Pennsylvania with the Carboniferous of those States. Thus we have data for the comparison of successive floras in the same region. In the Canadian region we have, indeed, in direct sequence, the floras of the Upper Silurian, the Lower, Middle, and Upper Erian, and the Lower, Middle, and Upper Carboniferous, all more or less distinct from each other, and affording an admirable series for comparison in a region whose geographical features are very broadly marked. All these floras are composed in great part of similar types, and probably do not indicate very dissimilar general physical conditions, but they are separated from each other by the great subsidences of the Corniferous limestone and the Lower Carboniferous limestone, and by the local but intense subterranean action which has altered and disturbed the Erian beds toward the close of that period. Still, these changes were not universal. The Corniferous limestone is absent in Gaspé, and probably in New Brunswick, where, consequently, the Erian flora could continue undisturbed during that long period. The Carboniferous limestone is absent from the slopes of the Appalachians in Pennsylvania, where a retreat may have been afforded to the Upper Erian and Lower Carboniferous floras. The disturbances at the close of the Erian were limited to those eastern regions where the great limestone-producing subsidences were unfelt, and, on the other hand, are absent in Ohio, where the subsidences and marine conditions were almost at a maximum.

Bearing in mind these peculiarities of the area in question, we may now group in a tabular form the distinct specific types recognised in the Erian system, indicating, at the same time, those which are represented by identical species in the Carboniferous, those represented by similar species of the same general type, and those not represented at all. For example, *Calamites cannæformis* extends as a species into the Carboniferous; *Asterophyllites latifolia* does not so extend, but is represented by closely allied species of the same type; *Nematophyton* disappears altogether before we reach the Carboniferous.

Table of Erian and Carboniferous Specific Types.

Erian types. Represented in Carboniferous—	By identical types.	By related forms.	Erian types. Represented in Carboniferous—	By identical types.	By related forms.
1. Syringoxylon mirabile?			27. Cordaites Robbii		*
2. Nematoxylon			28. C. angustifolia		
3. Nematophyton			29. Archæopteris Jacksoni		
4. Aporoxylon			30. Aneimites obtusa		*
5. Ormoxylon			31. Platyphyllum Brownii		
6. Dadoxylon		*	32. Cyclopteris varia		*
7. Sigillaria Vanuxemii		*	33. C. obtusa		
8. S. palpebra		*	34. Neuropteris polymorpha		*
9. Didymophyllum			35. N. serrulata		*
10. Calamodendron		*	36. N. retorquata		*
11. Calamites transitionis	*		37. N. resecta		
12. C. cannæformis	*		38. Megalopteris Dawsoni		
13. Asterophyllites scutigera			39. Sphenopteris Hœninghausi	*	
14. A. latifolia		*	40. S. Harttii		*
15. Annularia laxa			41. Hymenophyllites curtilobus		
16. Sphenophyllum antiquum		*	42. H. obtusilobus		*
17. Cyclostigma			43. Alethopteris discrepans		*
18. Arthrostigma			44. Pecopteris serrulata		*
19. Lepidodendron Gaspianum		*	45. P. preciosa		
20. L. corrugatum	*		46. Trichomanites		*
21. Lycopodites Matthewi		*	47. Callipteris		*
22. L. Richardsoni			48. Cardiocarpum		*
23. Ptilophyton Vanuxemii			49. C. Crampii		
24. Lepidophloios antiquus		*	50. Antholithes		*
25. Psilophyton princeps			51. Trigonocarpum		*
26. P. robustius					

Of the above forms, fifty-one in all, found in the Erian of eastern America, all, except the last four, are certainly distinct specific types. Of these only four reappear in the Carboniferous under identical species, but no less than twenty-six reappear under representative or allied forms, some at least of which a derivationist might claim as modified descendants. On the other hand, nearly one half of the Devonian types are unknown in the Carboniferous, while there remain a very large number of Carboniferous types not accounted for by anything known in the Devonian. Further, a very poor flora, including only two or three types, is the predecessor of the Erian flora in the Upper Silurian, and the flora again becomes poor in the Upper Devonian and Lower Carboniferous. Every new species discovered must more or less modify the above statements, and the whole Erian flora of America, as well as the Carboniferous, requires a thorough comparison with that of Europe before general conclusions can be safely drawn. In the mean time I may indicate the direction in which the facts seem to point by the following general statements :

1. Some of the forms reckoned as specific in the Devonian and Carboniferous may be really derivative races. There are indications that such races may have originated in one or more of the following ways : (1) By a natural tendency in synthetic types to become specialised in the direction of one or other of their constituent elements. In this way such plants as *Arthrostigma* and *Psilophyton* may have assumed new varietal forms. (2) By embryonic retardation or acceleration,* whereby certain species may have had their maturity advanced or postponed, thus giving them various grades of perfection in reproduction and complexity of structure. The fact that so many Erian and Carboniferous plants seem to be on the con-

---

* In the manner illustrated by Hyatt and Cope.

fines of the groups of Acrogens and Gymnosperms may be supposed favourable to such exchanges. (3) The contraction and breaking up of floras, as occurred in the Middle Erian and Lower Carboniferous, may have been eminently favourable to the production of such varietal forms as would result from what has been called the "struggle for existence." (4) The elevation of a great expanse of new land at the close of the Middle Erian and the beginning of the coal period would, by permitting the extension of species over wide areas and fertile soils, and by removing the pressure previously existing, be eminently favourable to the production of new, and especially of improved, varieties.

2. Whatever importance we may attach to the above supposed causes of change, we still require to account for the origin of our specific types. This may forever elude our observation, but we may at least hope to ascertain the external conditions favourable to their production. In order to attain even to this it will be necessary to inquire critically, with reference to every acknowledged species, what its claims to distinctness are, so that we may be enabled to distinguish specific types from mere varieties. Having attained to some certainty in this, we may be prepared to inquire whether the conditions favourable to the appearance of new varieties were also those favourable to the creation of new types, or the reverse—whether these conditions were those of compression or expansion, or to what extent the appearance of new types may be independent of any external conditions, other than those absolutely necessary for their existence. I am not without hope that the further study of fossil plants may enable us thus to approach to a comprehension of the laws of the creation, as distinguished from those of the continued existence of species.

3. In the present state of our knowledge we have no good ground either to limit the number of specific types

beyond what a fair study of our material may warrant, or to infer that such primitive types must necessarily have been of low grade, or that progress in varietal forms has always been upward. The occurrence of such an advanced and specialised type as that of *Dadoxylon* in the Middle Devonian should guard us against these errors. The creative process may have been applicable to the highest as well as to the lowest forms, and subsequent deviations must have included degradation as well as elevation. I can conceive nothing more unreasonable than the statement sometimes made that it is illogical or even absurd to suppose that highly organised beings could have been produced except by derivation from previously existing organisms. This is begging the whole question at issue, depriving science of a noble department of inquiry on which it has as yet barely entered, and anticipating by unwarranted assertions conclusions which may perhaps suddenly dawn upon us through the inspiration of some great intellect, or may for generations to come baffle the united exertions of all the earnest promoters of natural science. Our present attitude should not be that of dogmatists, but that of patient workers content to labour for a harvest of grand generalisations which may not come till we have passed away, but which, if we are earnest and true to Nature and its Creator, may reward even some of us.

Within the human period great changes of distribution of plants have occurred, chiefly through the agency of man himself, and we have had ample evidence that plants are able to establish themselves and prosper in climates and conditions to which unaided they could not have transported themselves, as, for instance, in the case of European weeds naturalised in Australia and New Zealand. There is, however, no reason to believe that any specific change has occurred to any plant within the Pleistocene or modern period.

In a recent address, delivered to the biological section of the British Association, Mr. Carruthers has discussed this question, and has shown that the earliest vegetable specimens described by Dr. Schweinfurth from the Egyptian tombs present no appearance of change. This fact appears also in the leaves and other organs of plants preserved in the nodules in the Pleistocene clays of the Ottawa, and in specimens of similar age found in various places in Britain and the continent of Europe.*

The difficulties attending the ordinary theories of evolution as applied to plants have been well set forth by the same able botanist in his " Presidential Address to the Geological Association in 1877," a paper which deserves careful study. One of his illustrations is that ancient willow, *Salix polaris,* referred to in a previous chapter, which now lives in the arctic regions, and is found fossil in the Pleistocene beds at Cromer and at Bovey Tracey.

He notes the fact that the genus *Salix* is a very variable one, including 19 subgeneric groups and 160 species, with no less than 222 varieties and 70 hybrids. *Salix polaris* belongs to a subgeneric group containing 29 species, which are arranged in four sections, that to which *S. polaris* belongs containing six species. Now it is easy to construct a theoretical phylogeny of the derivation of the willows from a supposed ancestral source, but when we take our little *S. polaris* we find that this one twig of our ancestral tree takes us back without change to the Glacial period. The six species would take us still farther, and the sections, subgenera, and genus at the same rate would require an incalculable amount of past time. He concludes the inquiry in the following terms :

---

* " Proceedings British Association," 1886, " Pleistocene Plants of Canada," Canadian Naturalist, 1866.

"But when we have reached the branch representing the generic form we have made but little progress in the phylogenesis of *Salix*. With *Populus* this genus forms a small order, Salicineæ. The two genera are closely allied, yet separated by well-marked characters; it is not, however, difficult to conceive of both having sprung from a generalised form. But there is no record of such a form. The two genera appear together among the earliest known dicotyledons, the willows being represented by six and the poplars by nine species. The ordinal form, if it ever existed, must necessarily be much older than the period of the Upper Cretaceous rocks, that is, than the period to which the earliest known dicotyledons belong.

"The Salicineæ are related to five other natural orders, in all of which the apetalous flowers are arranged in catkins. These different though allied orders must be led up by small modifications to a generalised amentiferous type, and thereafter the various groups of apetalous plants by innumerable eliminations of differentiating characters until the primitive form of the apetalous plant is reached. Beyond this the uncurbed imagination will have more active work in bridging over the gap between Angiosperms and Gymnosperms, in finding the intermediate forms that led up to the vascular cryptogams, and on through the cellular plants to the primordial germ. Every step in this phylogenetic tree must be imagined. The earliest dicotyledon takes us not a step farther back in the phylogenetic history of *Salix* than that supplied by existing vegetation. All beyond the testimony of our living willows is pure imagination, unsupported by a single fact. So that here, also, the evidence is against evolution, and there is none in favour of it."

It is easy to see that similar difficulties beset every attempt to trace the development of plants on the principle of slow and gradual evolution, and we are driven back on the theory of periods of rapid origin, as we have already seen suggested by Saporta in the case of the Cretaceous dicotyledons.   Such abrupt and plentiful introduction of species over large areas at the same time, by whatever cause effected—and we are at present quite ignorant of any secondary causes—becomes in effect something not unlike the old and familiar idea of creation.   Science must indeed always be baffled by questions of ultimate origin, and, however far it may be able to trace the chain of secondary causation and development, must at length find itself in the presence of the great Creative Mind, who is "before all things and in whom all things consist."

APPENDIX.

I.—COMPARATIVE VIEW OF THE SUCCESSIVE PALÆO-ZOIC FLORAS OF NORTHEASTERN AMERICA AND GREAT BRITAIN.

In eastern Canada there is a very complete series of fossil plants, extending from the Silurian to the Permian, and intermediate in its species between the floras of interior America and of Europe. I may use this succession, mainly worked out by myself,* to summarise the various Palæozoic floras and sub-floras, in order to give a condensed view of this portion of the history of the vegetable kingdom, and to direct attention to the important fact, too often overlooked, that there is a definite succession of fossil plants as well as of animals, and that this is important as a means of determining geological horizons. A British list for comparison has been kindly prepared for me by Mr. R. Kidston, F. G. S. For lists referring to the western and southern portions of America, I may refer to the reports of Lesquereux and Fontaine and White.†

In this connection I am reminded, by an excellent little paper of M. Zeiller, ‡ on Carboniferous plants from the region of the Zambesi, in Africa, that the flora which in the Carboniferous period extended over the temperate portions of the northern hemisphere and far into the arctic, also passed across the equator and prevailed in the southern hemisphere. Of eleven species brought from the Zambesi by M. Lapierre and examined by M. Zeiller, all were identical with Euro-

---

* "Acadian Geology," "Reports on Fossil Plants of Canada," Geological Survey of Canada.

† "Geological Surveys of Pennsylvania, Ohio, and Illinois."

‡ Paris, 1883.

T

pean species of the upper coal-formation, and the same fact has been observed in the coal flora of the Cape Colony.* These facts bear testimony to the remarkable uniformity of climate and vegetation in the coal period, and I perfectly agree with Zeiller that they show, when taken in connection with other parallelisms in fossils, an actual contemporaneousness of the coal flora over the whole world.

1. CARBONIFEROUS FLORA.

(1) *Permo-Carboniferous Sub-Flora:*

This occurs in the upper member of the Carboniferous system of Nova Scotia and Prince Edward Island, originally named by the writer the Newer Coal-formation, and more ´recently the Permo-Carboniferous, and the upper beds of which may not improbably be contemporaneous with the Lower Permian or Lower Dyas of Europe. In this formation there is a predominance of red sandstones and shales, and it contains no productive beds of coal. Its fossil plants are for the most part of species found in the Middle or Productive Coal-formation, but are less numerous, and there are a few new forms akin to those of the European Permian. The most characteristic species of the upper portion of the formation, which has the most decidedly Permian aspect, are the following:

Dadoxylon materiarium, Dawson.
* *Walchia (Araucarites) robusta*, Dn.
* *W. (A.) gracilis*, Dn.
* *W. imbricatula*, Dn.
Calamites Suckovii, Brongt.
C. Cistii, Brongt.
* *C. gigas*, Brongt.
Neuropteris rarinervis, Bunbury.
Alethopteris nervosa, Brongt.
Pecopteris arborescens, Brongt.
* *P. rigida*, Dn.
P. oreopteroides, Brongt.
* *Cordaites simplex*, Dn.

Of these species, those marked with an asterisk have not yet been found in the middle or lower members of the Carboniferous system. They will be found described, and several of them figured, in my " Report on the Geology of Prince Edward Island." † The others are

---

* Grey, " Journal of the Geological Society," vol. xxvii.
† 1871.

common and widely diffused Carboniferous species, some of which have extended to the Permian period in Europe as well. From the upper beds, characterised by these and a few other species, there is a gradual passage downward into the productive coal-measures, and a gradually increasing number of true coal-formation species.

It is worthy of remark here that the association in the Permo-Carboniferous of numerous trunks of *Dadoxylon* with the branches of *Walchia* and with fruits of the character of *Trigonocarpa*, seems to show that these were parts of one and the same plant.

This formation represents the Upper Barren Measures of West Virginia, which are well described by Fontaine and White,[*] and the reasons which these authors adduce for considering the latter equivalent to the European Permian will apply to the more northern and eastern deposits as well, though these have afforded fewer species of plants, and are apparently less fully developed.

(2) *Coal-formation Sub-Flora:*

The Middle or Productive Coal-formation, containing all the beds of coal which are mined in Nova Scotia and Cape Breton, is the headquarters of the Carboniferous flora. From this formation I have catalogued [†] one hundred and thirty-five species of plants; but, as several of these are founded on imperfect specimens, the number of actual species may be estimated at one hundred and twenty. Of these more than one half are species common to Europe and America. No less than nineteen species are *Sigillariæ*, and about the same number are *Lepidodendra.* About fifty are ferns and thirteen are *Calamites, Asterophyllites,* and *Sphenophylla.* The great abundance and number of species of *Sigillariæ, Lepidodendra,* and ferns are characteristic of this sub-flora; and among the ferns certain species of *Neuropteris, Pecopteris, Alethopteris,* and *Sphenopteris* greatly preponderate.

These beds are the equivalents of the Middle Coal-measures, or Productive Coal-measures of Pennsylvania, Ohio, &c., and of the coal-formation proper of various European countries. Very many of the species are common to Nova Scotia and Pennsylvania; but in proceeding westward the number of identical species seems to diminish.

---

[*] " Report on the Permian Flora of Western Virginia and South Pennsylvania," 1880.

[†] "Acadian Geology," and " Report on Flora of Lower Carboniferous," 1873.

(3) *The Millstone Grit Sub-Flora:*

In this formation the abundance of plants and the number of species are greatly diminished.* Trunks of coniferous trees of the species *Dadoxylon Acadianum,* having wide wood-cells with three or more series of discs and complex medullary rays, become characteristic. *Calamites undulatum* is abundant and seems to replace *C. Suckovii,* though *C. cannæformis* and *C. cistii* continue. *Sigillariæ* become very rare, and the species of *Lepidodendron* are few, and mostly those with large leaf-bases. *Lepidophloios* still continues, and *Cordaites* abounds in some beds. The ferns are greatly reduced, though a few characteristic coal-formation species occur, and the genus *Cardiopteris* appears. Beds of coal are rare in this formation; but where they occur there is in connection with them a remarkable anticipation of the rich coal-formation flora, which would thus seem to have existed locally in the Millstone Grit period, but to have found itself limited by generally unfavorable conditions. In America, as in Europe, it is in the north that this earlier development of the coal-flora occurs, while in the south there is a lingering of old forms in the newer beds. In Newfoundland and Cape Breton, for instance, as well as in Scotland, productive coal-beds and a greater variety of species of plants occur in this formation.

The following would appear to be the equivalents of this formation, in flora and geological position:

1. The Seral Conglomerate of Rogers in Pennsylvania, &c.

2. The Lower Coal-formation Conglomerate and Chester groups of Illinois (Worthen).

3. The Lower Carboniferous Sandstone of Kentucky, Alabama, and Virginia.

4. The Millstone Grit and Yoredale rocks of northern England, and the Culmiferous of Devonshire.

5. The Moor rock and Lower Coal-measures of Scotland.

6. Flagstones and Lower Shales of the south of Ireland, and Millstone Grit of the north of Ireland.

7. The Jüngste Grauwacke of the Hartz, Saxony, and Silesia.

(4) *The Carboniferous Limestone Series:*

This affords few fossil plants in eastern America, and in so far as known they are similar to those of the next group. In Scotland it is richer in plants, but, according to Mr. Kidston, these are largely

---

* "Report on Fossil Plants of the Lower Carboniferous and Millstone Grit of Canada," 1873.

similar to those of the underlying beds, though with some species which extend upward into the Millstone Grit. In Scotland the alga named *Spirophyton* and *Archæocalamites radiatus*—which in America are Erian—appear in this formation.

(5) *The Lower Carboniferous Sub-Flora :*

This group of plants is best seen in the shales of the Horton series, under the Lower Carboniferous marine limestones. It is small and peculiar. The most characteristic species are the following :

Dadoxylon (Palæoxylon) antiquius, Dn.—A species with large medullary rays of three or more series of cells.

Lepidodendron corrugatum, Dn.—A species closely allied to *L. Veltheimianum* of Europe, and which is its American representative. This is perhaps the most characteristic plant of the formation. It is very abundant, and presents very protean appearances, in its old stems, branches, twigs, and *Knorria* forms. It had well-characterised stigmaria roots, and constitutes the oldest erect forest known in Nova Scotia.

Lepidodendron tetragonum, Sternberg.

L. obovatum, Sternb.

L. aculeatum, Sternb.

L. dichotomum, Sternb.

The four species last mentioned are comparatively rare, and the specimens are usually too imperfect to render their identification certain, but Lepidodendra are especially characteristic trees of this horizon.

Cyclopteris (Aneimites) Acadica, Dn.—A very characteristic fern, allied in the form of its fronds to *C. tenuifolia* of Goeppert, to *C. nana* of Eichwald, and to *Adiantites antiquus* of Stur. Its fructification, however, is nearer to that of *Aneimia* than to that of *Adiantum*.

Ferns of the genera *Cardiopteris* and *Hymenophyllites* also occur, though rarely.

Ptilophyton plumula, Dn.—This is the latest appearance of this Erian genus, which also occurs in the Lower Carboniferous of Europe and of the United States.

Cordaites borassifolia, Brongt.

On the whole, this small flora is markedly distinct from that of the Millstone Grit and true coal-formation, from which it is separated by the great length of time required for the deposition of the marine limestones and their associated beds, in which no land-plants have been found; nor is this gap filled up by the conglomerates and coarse arenaceous beds which, as I have explained in "Acadian Geology," in some localities take the place of the limestones, as they do also in the Appalachian region farther south.

The palæobotanical and strategraphical equivalents of this series abroad would seem to be the following:

1. The Vespertine group of Rogers in Pennsylvania.
2. The Kinderhook group of Worthen in Illinois.
3. The Marshall group of Winchell in Michigan.
4. The Waverley sandstone (in part) of Ohio.
5. The Lower or False Coal-measures of Virginia.
6. The Calciferous sandstones of McLaren, or Tweedian group of Tate in Scotland.
7. The Lower Carboniferous slate and Coomhala grits of Jukes in Ireland.
8. The Culm and Culm Grauwacke of Germany.
9. The Graywacke or Lower Coal-measures of the Vosges, as described by Schimper.
10. The Older Coal-formation of the Ural, as described by Eichwald.
11. The so-called "Ursa Stage" of Heer includes this, but he has united it with Devonian beds, so that the name cannot be used except for the local development of these beds at Bear Island, Spitzbergen. The Carboniferous plants of arctic America, Melville Island, &c., as well as those of Spitzbergen, appear all to be Lower Carboniferous.*

All of the above groups of rocks are characterised by the prevalence of *Lepidodendra* of the type of *L. corrugatum, L. Veltheimianum*, and *L. Glincanum;* pines of the sub-genus *Pitus* of Witham, *Palæoxylon* of Brongniart, and peculiar ferns of the genera *Cyclopteris, Cardiopteris, Triphyllopteris,* and *Sphenopteris.* In all the regions above referred to they form the natural base of the great Carboniferous system.

In Virginia, according to Fontaine and White, types, such as *Archæopteris,* which in the north are Upper Erian, occur in this group. Unless there have been some errors in fixing the lower limit of the Vespertine, this would indicate a longer continuance of old forms in the south.

---

* "Notes on Geological Map of the Northern Portion of the Dominion of Canada," by Dr. G. M. Dawson, 1887.

2. ERIAN FLORA.

(1) *Upper Erian Sub-Flora:*

This corresponds to the Catskill and Chemung of the New York series, and to the Upper Devonian of Europe.

The flora of this formation, which consists mostly of sandstones, is not rich. Its most distinctive species on both sides of the Atlantic seem to be the ferns of the genus *Archæopteris*, along with species referred to the genus *Cyclopteris*, but which, in so far as their barren fronds are concerned, for the most part resemble *Archæopteris*.

The characteristic American species are *Archæopteris Jacksoni*, *A. Rogersi*, and *A. Gaspiensis*. *Cyclopteris obtusa* and *C. (Platyphyllum) Brownii* are also very characteristic species. In Europe, *Archæopteris Hibernica* is a prevalent species.

Leptophleum rhombicum and fragments of *Psilophyton* are also found in the Upper Erian. There is evidence of the existence of vast numbers of *Rhizocarps* in this period, in the deposits of spore-cases (*Sporangites Huronensis*) in the shales of Kettle Point, Lake Huron; and in deposits of similar character in Ohio and elsewhere in the West.

The Upper Erian flora is thus very distinct from that of the Lower Carboniferous, and the unconformable relation of the beds in the Northeast may perhaps indicate a considerable lapse of time. Still, even in localities where there appears to be a transition from the Carboniferous into the Devonian, as in the Western States and in Ireland, the characteristic flora of each formation may be distinguished, though, as already stated, there is apparently some mixture in the South.

(2) *Middle Erian Sub-Flora:*

Both in Canada and the United States that part of the great Erian system which may be regarded as its middle division, the Hamilton and Marcellus shales of New York, the Cordaites shales of St. John, New Brunswick, and the middle shales and sandstones of the Gaspé series, presents conditions more favourable to the abundant growth of land-plants than either the upper or lower member. In the St. John beds, in particular, there is a rich fern flora, comparable with that of the coal-formation, and numerous stipes of ferns and trunks of tree-ferns have been found in the Hamilton and Corniferous series in the West, as well as trunks of *Dadoxylon*. It is, however, distinguished by a prevalence of small and delicate species, and by such forms as *Hymenophyllites* and the smaller Sphenopterids, and also by some peculiar ferns, as *Archæopteris* and *Megalopteris*.

In addition to ferns, it has small *Lepidodendra*, of which *L. Gaspianum* is the chief. *Calamiteæ* occur, *Archæocalamites radiatus* being the dominant species. This plant, which in Europe appears to reach up into the Lower Carboniferous, is so far strictly Erian in northeast America. *Sigillariæ* scarcely appear, but *Cordaites* is abundant, and the earliest known species of *Dadoxylon* appear, while the *Psilophyton*, so characteristic of the Lower Erian, still continues, and the remarkable aquatic plants of the genus *Ptilophyton* are locally abundant.

(3) *Lower Erian Sub-Flora :*

This belongs to the Lower Devonian sandstones and shales, and is best seen in that formation at Gaspé and the Bay des Chaleurs. It is equivalent to the Oriskany sandstone, so far as its animal fossils and mineral character are concerned. It is characterised by the absence of true ferns, *Calamites* and *Sigillariæ*, and by the presence of such forms as *Psilophyton, Arthrostigma, Leptophleum,* and *Nematophyton*. *Lepidodendron Gaspianum* and *Leptophleum* already occur, though not nearly so abundant as *Psilophyton*.

The Lower Erian plants have an antique and generalised aspect which would lead us to infer that they are near the beginning of the land-flora, or perhaps in part belong to the close of an earlier flora still in great part unknown; and few indications of land-plants have been found earlier.

At Campbellton and Scaumenac Bay, on the Bay des Chaleurs, fossil fishes of genera characteristic of the Lower and Upper Devonian horizons respectively, occur in association with fossil plants of these horizons, and have been described by Mr. Whiteaves.*

It is interesting to note that, as Fontaine and White have observed, certain forms which are Erian in the northeast are found in the Lower members of the Carboniferous in West Virginia, indicating the southward march of species in these periods.

3. The Silurian Flora and still Earlier Indications of Plants.

In the upper beds of the Silurian, those of the Helderberg series, we still find *Psilophyton* and *Nematophyton ;* but below these we know no land-plants in Canada. In the United States, Lesquereux and Claypole have described remains which may indicate the existence of lycopodiaceous and annularian types as far back as the be-

---

* "Transactions of the Royal Society of Canada."

ginning of the Upper Silurian, or even as low as the Hudson River group, and Hicks has found *Nematophyton* and *Psilophyton* in beds about as old in Wales, along with the uncertain stems named *Berwynia*. In the Lower Silurian the *Protannularia* of the Skiddaw series in England may represent a land-plant, but this is uncertain, and no similar species has been found in Canada.

The Cambrian rocks are so far barren of land-plants; the so-called *Eophyton* being evidently nothing but markings, probably produced by crustaceans and other aquatic animals. In the still older Laurentian the abundant beds of graphite probably indicate the existence of plants, but whether aquatic or terrestrial it is impossible to decide at present.

It would thus appear that our certain knowledge of land-vegetation begins with the Upper Silurian or the Silurio-Cambrian, and that its earliest forms were Acrogens allied to Lycopods, and prototypal trees, forerunners of the Acrogens or the gymnosperms. In the Lower Devonian little advance is made. In the Middle Devonian this meagre flora had been replaced by one rivalling that of the Carboniferous, and including pines, tree-ferns, and arboreal forms of Lycopods and of equisetaceous plants, as well as numerous herbaceous plants. At the close of the Erian the flora again became meagre, and continued so in the Lower Carboniferous. It again became rich and varied in the Middle Carboniferous, to decay in the succeeding Permian.

II.—HEER'S LATEST RESULTS IN THE GREENLAND FLORA.

A VERY valuable report of Prof. Steenstrup, published in Copenhagen in 1883, the year in which Heer died, contains the results of his last work on the Greenland plants, and is so important that a summary of its contents will be interesting to all students of fossil botany or of the vicissitudes of climate which the earth has undergone.[*]

The plant-bearing beds of Greenland are as follows, in ascending order:

1. CRETACEOUS.

1. The *Komé* series, of black shales resting on the Laurentian gneiss. These beds are found at various other localities, but the

---

[*] Meddelclser om Gronland, Hefte V., Copenhagen, 1883.

name above given is that by which they are generally kn⟨ vn. Their
flora is limited to ferns, cycads, conifers, and a few endogens, with
only *Populus primæva* to represent the dicotyledons. These beds
are regarded as Lower Cretaceous (Urgonian), but the animal fossils
would seem to give them a rather higher position. They may be
regarded as equivalent to the Kootanie and Queen Charlotte beds in
Canada, and the Potomac series in Virginia.

2. The *Atané* series. These also are black shales with dark-
coloured sandstones. They are best exposed at Upernavik and
Waigat. Here dicotyledonous leaves abound, amounting to ninety
species, or more than half the whole number of species found.
The fossil plants resemble those of the Dakota series of the United
States and the Dunvegan series of Canada, and the animal fossils
indicate the horizon of the Fort Pierre or its lower part. They may
be regarded as representing the lower part of the Upper Cretaceous.
The genera *Populus, Myrica, Quercus, Ficus, Platanus, Sassafras,
Laurus, Magnolia,* and *Liriodendron* are among those represented
in these beds, and the peculiar genera *Macclintockia* and *Credneria*
are characteristic. The genus *Pinus* is represented by five species,
Sequoia by five, and *Salisburia* by two, with three of the allied
genus *Baiera.* There are many ferns and cycads.

3. The *Patoot* series. These are yellow and red shales, which
seem to owe their colour to the spontaneous combustion of pyritous
lignite, in the manner observed on the South Saskatchewan and the
Mackenzie rivers. Their age is probably about that of the Fox-Hill
group or Senonian, and the Upper Cretaceous of Vancouver Island,
and they afford a large proportion of dicotyledonous leaves. The
genera of dicotyledons are not dissimilar from those of Atané, but
we now recognise *Betula* and *Alnus, Comptonia, Planera, Sapo-
tacites, Fraxinus, Viburnum, Cornus, Acer, Celastrus, Paliurus,
Ceanothus, Zizyphus,* and *Cratægus* as new genera of modern aspect.

On the whole there have been found in all these beds 335 species,
belonging to 60 families, of which 36 are dicotyledonous, and represent all the leading types of arborescent dicotyledons of the temperate latitudes. The flora is a warm temperate one, with some remarkable mixtures of sub-tropical forms, among which perhaps the
most remarkable are *Kaidocarpum* referred to the *Pandaneæ,* and
such exogens as *Ficus* and *Cinnamomum.*

2. TERTIARY.

4. The *Unartok* series. This is believed to be Eocene. It consists of sandstone, which appears on the shores of Disco Island, and

possibly at some other places on the coast. The beds rest directly and apparently conformably on the Upper Cretaceous, and have afforded only eleven species of plants. *Magnolia* is represented by two species, *Laurus* by two, *Platanus* by two, and one of these said to be identical with a species found by Lesquereux in the Laramie,[*] *Viburnum, Juglans, Quercus*, each by one species; the ubiquitous Sequoias by *S. Langsdorffii*. This is pretty clearly a Lower Laramie flora.

5. The *Atanekerdluk* series, consisting of shaly beds, with limestone intercalated between great sheets of basalt, much like the Eocene of Antrim and the Hebrides. These beds have yielded 187 species, principally in bands and concretions of siderite, and often in a good state of preservation. They are referred to the Lower Miocene, but, as explained in the text, the flora is more nearly akin to that of the Eocene of Europe and the Laramie of America. The animal fossils are chiefly fresh-water shells. *Onoclea sensibilis*, several conifers, as *Taxites Olriki, Taxodium distichum, Glyptostrobus Europæus*, and *Sequoia Langsdorffii*, and 42 of the dicotyledons are recognised as found also in American localities. Of these, a large proportion of the more common species occur in the Laramie of the Mackenzie River and elsewhere in northwest Canada, and in the western United States. It is quite likely also that several species regarded as distinct may prove to be identical.

It would seem that throughout the whole thickness of these Tertiary beds the flora is similar, so that it is probable it belongs altogether to the Eocene rather than to the Miocene.

No indication has been observed of any period of cold intervening between the Lower Cretaceous and the top of the Tertiary deposits, so that, in all the vast period which these formations represent, the climate of Greenland would seem to have been temperate. There is, however, as is the case farther south, evidence of a gradual diminution of temperature. In the Lower Cretaceous the probable mean annual temperature in latitude 71° north is stated as 21° to 22° centigrade, while in the early Tertiary it is estimated at 12° centigrade. Such temperatures, ranging from 71° to 53° of Fahrenheit, represent a marvellously warm climate for so high a latitude. In point of fact, however, the evidence of warm climates in the arctic regions, in the Palæozoic as well as in the Mesozoic and early Tertiary, should perhaps lead us to conclude that, relatively to the whole of geological time, the present arctic climate is unusually severe, and

---

[*] *Viburnum marginatum* of Lesquereux.

that a temperate climate in the arctic regions has throughout geological time been the rule rather than the exception.

III.—MINERALISATION OF FOSSIL PLANTS.

THE state of preservation of fossil plants has been referred to incidentally in several places in the text; but the following more definite statements may be of service to the reader.

I. Organic remains imbedded in aqueous deposits may occur in an unchanged condition, or only more or less altered by decay. This is often the case with such enduring substances as bark and wood, and even with leaves, which appear as thin carbonaceous films when the layers containing them are split open. In the more recent deposits such remains occur little modified, or perhaps only slightly changed by partial decay of their more perishable parts. In the older formations, however, they are usually found in a more or less altered condition, in which their original substance has been wholly or in part changed into coaly, or bituminous, or anthracitic or graphitic matter, so that leaves are sometimes represented by stains of graphite, as if drawn on stone with a lead-pencil. Yet even in this case some portion of the original substance remains, and without any introduction of foreign material.

II. On the other hand, such remains are often mineralised by the filling of their pores or the replacement of their tissues with mineral matter, so that they become hard and stony, and sometimes retain little or nothing of their original substance. The more important of these changes, in so far as they affect fossil plants, may be arranged under the following heads:

(a) *Infiltration* of mineral matter which has penetrated the pores of the fossil in a state of solution. Thus the pores of fossil wood are often filled with calcite, quartz, oxide of iron, or sulphide of iron, while the woody walls of the cells and vessels remain in a carbonised state, or converted into coaly matter. When wood is preserved in this way it has a hard and stony aspect; but we can sometimes dissolve away the mineral matter, and restore the vegetable tissue to a condition resembling that before mineralisation. This is especially the case when calcite is the mineralising substance. We sometimes find, on microscopic examination, that even cavities so small as those of vegetable cells and vessels have been filled with successive coats of different kinds of mineral matter.

(b) Organic matters may be entirely *replaced* by mineral substances. In this case the cavities and pores have been first filled, and then—the walls or solid parts being removed by decay or solution—mineral matter, either similar to that filling the cavities, or differing in colour or composition, has been introduced. Silicified wood often occurs in this condition. In the case of silicified wood, it sometimes happens that the cavities of the fibers have been filled with silica, and the wood has been afterward removed by decay, leaving the casts of the tubular fibers as a loose filamentous substance. Some of the Tertiary coniferous woods of California are in this state, and look like asbestus, though they show the minute markings of the tissue under the microscope. In the case of silicified or agatized woods, it would seem that the production of carbon dioxide from the decaying wood has caused the deposition of silica in its place, from alkaline solutions of that substance, and thus the carbon has been replaced, atom by atom, by silicon, until the whole mass has been silicified, yet retaining perfectly its structure.

(c) The cavities left by fossils which have decayed may be filled with clay, sand, or other foreign matter, and this, becoming subsequently hardened into stone, may constitute a *cast* of the fossils. Trunks of trees, roots, &c., are often preserved in this way, appearing as stony casts, often with the outer bark of the plant forming a carbonaceous coating on their surfaces. In connection with this state may be mentioned that in which, the wood having decayed, an entire trunk has been flattened so as to appear merely as a compressed film of bark, yet retaining its markings; and that in which the whole of the vegetable matter having been removed, a mere impression of the form remains.

Fossils preserved in either of the modes, (a) or (b), usually show more or less of their minute structures under the microscope. These may be observed :—(1) By breaking off small splinters or flakes and examining them, either as opaque or as transparent objects. (2) By treating the material with acids, so as to dissolve out the mineral matters, or portions of them. This method is especially applicable to fossil woods mineralised with calcite or pyrite. (3) By grinding thin sections. These are first polished on one face on a coarse stone or emery hone, and then on a fine hone, then attached by the polished face to glass slips with a transparent cement or Canada balsam, and ground on the opposite face until they become so thin as to be translucent. In most cities there are lapidaries who prepare slices of this kind; but the amateur can readily acquire the art by a little practice, and the necessary appliances can be obtained through dealers in minerals or in microscopic materials. Very convenient cutting and polishing machines, some of them quite small and portable, are

now made for the use of amateurs. In the case of exogenous woods,
three sections are necessary to exhibit the whole of the structures.
One of these should be transverse and two longitudinal, the latter in
radial and tangential planes.

IV.—GENERAL WORKS ON PALÆOBOTANY.

IN the text frequent reference has been made to special memoirs
and reports on the fossil plants of particular regions or formations.
There are, however, some general books, useful to students, which
may be mentioned here. Perhaps the most important is Schimper's
" Traité de Paléontologie Végétale." Very useful information is
also contained in Renault's " Cours de Botanique Fossile," and in
Balfour's " Introduction to Palæontological Botany," and Nicholson's " Palæontology." Unger's " Genera et Species," Brongniart's
" Histoire des Végétaux Fossiles," and Lindley and Hutton's " Fossil
Flora," are older though very valuable works. Williamson's " Memoirs," in the " Philosophical Transactions," have greatly advanced
our knowledge of the structures of Palæozoic plants. Lastly, the
" Palæophytology " of Schenk, now in course of publication in German and French, in connection with Zittel's " Palæontology," is an
important addition to manuals of the subject.

INDEX.

THE END.

PRINTED BY WILLIAM CLOWES AND SONS, LIMITED,
LONDON AND BECCLES.

A LIST OF

KEGAN PAUL, TRENCH & CO.'S

PUBLICATIONS.

11,87.

1, *Paternoster Square,*
London.

A LIST OF
KEGAN PAUL, TRENCH & CO.'S
PUBLICATIONS.

CONTENTS.

GENERAL LITERATURE.

A. K. H. B. — From a Quiet Place. A Volume of Sermons. Crown 8vo, 5s.

ALEXANDER, William, D.D., Bishop of Derry.—The Great Question, and other Sermons. Crown 8vo, 6s.

ALLIES, T. W., M.A.—Per Crucem ad Lucem. The Result of a Life. 2 vols. Demy 8vo, 25s.

A Life's Decision. Crown 8vo, 7s. 6d.

AMHERST, Rev. W. J.—The History of Catholic Emancipation and the Progress of the Catholic Church in the British Isles (chiefly in England) from 1771-1820. 2 vols. Demy 8vo, 24s.

AMOS, Professor Sheldon.—The History and Principles of the Civil Law of Rome. An aid to the Study of Scientific and Comparative Jurisprudence. Demy 8vo, 16s.

Ancient and Modern Britons. A Retrospect. 2 vols. Demy 8vo, 24s.

ARISTOTLE.—The Nicomachean Ethics of Aristotle. Translated by F. H. Peters, M.A. Third Edition. Crown 8vo, 6s.

AUBERTIN, J. J.—A Flight to Mexico. With 7 full-page Illustrations and a Railway Map of Mexico. Crown 8vo, 7s. 6d.

AUBERTIN, J. J.—continued.

Six Months in Cape Colony and Natal. With Illustrations and Map. Crown 8vo, 6*s.*

Aucassin and Nicolette. Edited in Old │French and rendered in Modern English by F. W. BOURDILLON. Fcap 8vo, 7*s.* 6*d.*

AUCHMUTY, A. C.—**Dives and Pauper, and other Sermons.** Crown 8vo, 3*s.* 6*d.*

AZARIUS, Brother.—**Aristotle and the Christian. Church.** Small crown 8vo, 3*s.* 6*d.*

BADGER, George Percy, D.C.L.—**An English-Arabic Lexicon.** In which the equivalent for English Words and Idiomatic Sentences are rendered into literary and colloquial Arabic. Royal 4to, 80*s.*

BAGEHOT, Walter. — **The English Constitution.** Fourth Edition. Crown 8vo, 7*s.* 6*d.*

Lombard Street. A Description of the Money Market. Eighth Edition. Crown 8vo, 7*s.* 6*d.*

Essays on Parliamentary Reform. Crown 8vo, 5*s.*

Some Articles on the Depreciation of Silver, and Topics connected with it. Demy 8vo, 5*s.*

BAGOT, Alan, C.E.—**Accidents in Mines:** their Causes and Prevention. Crown 8vo, 6*s.*

The Principles of Colliery Ventilation. Second Edition, greatly enlarged. Crown 8vo, 5*s.*

The Principles of Civil Engineering as applied to Agriculture and Estate Management. Crown 8vo, 7*s.* 6*d.*

BAIRD, Henry M.—**The Huguenots and Henry of Navarre.** 2 vols. With Maps. 8vo, 24*s.*

BALDWIN, Capt. J. H.—**The Large and Small Game of Bengal and the North-Western Provinces of India.** With 20 Illustrations. New and Cheaper Edition. Small 4to, 10*s.* 6*d.*

BALL, John, F.R.S.—**Notes of a Naturalist in South America.** With Map. Crown 8vo, 8*s.* 6*d.*

BALLIN, Ada S. and F. L.—**A Hebrew Grammar.** With Exercises selected from the Bible. Crown 8vo, 7*s.* 6*d.*

BARCLAY, Edgar.—**Mountain Life in Algeria.** With numerous Illustrations by Photogravure. Crown 4to, 16*s.*

BASU, K. P., M.A.—**Students' Mathematical Companion.** Containing problems in Arithmetic, Algebra, Geometry, and Mensuration, for Students of the Indian Universities. Crown 8vo, 6*s.*

BAUR, Ferdinand, Dr. Ph.—A Philological Introduction to Greek and Latin for Students. Translated and adapted from the German, by C. KEGAN PAUL, M.A., and E. D. STONE, M.A. Third Edition. Crown 8vo, 6s.

BAYLY, Capt. George.—Sea Life Sixty Years Ago. A Record of Adventures which led up to the Discovery of the Relics of the long-missing Expedition commanded by the Comte de la Perouse. Crown 8vo, 3s. 6d.

BENSON; A. C.—William Laud, sometime Archbishop of Canterbury. A Study. With Portrait. Crown 8vo, 6s.

BIRD, Charles, F.G.S.—Higher Education in Germany and England. Small crown 8vo, 2s. 6d.

Birth and Growth of Religion. A Book for Workers. Crown 8vo, cloth, 2s. ; paper covers, 1s.

BLACKBURN, Mrs. Hugh.—Bible Beasts and Birds. 22 Illustrations of Scripture photographed from the Original. 4to, 42s.

BLECKLY, Henry.—Socrates and the Athenians : An Apology. Crown 8vo, 2s. 6d.

BLOOMFIELD, The Lady.—Reminiscences of Court and Diplomatic Life. New and Cheaper Edition. With Frontispiece. Crown 8vo, 6s.

BLUNT, The Ven. Archdeacon.—The Divine Patriot, and other Sermons. Preached in Scarborough and in Cannes. New and Cheaper Edition. Crown 8vo, 4s. 6d.

BLUNT, Wilfrid S.—The Future of Islam. Crown 8vo, 6s.

Ideas about India. Crown 8vo. Cloth, 6s.

BODDY, Alexander A.—To Kairwân the Holy. Scenes in Muhammedan Africa. With Route Map, and Eight Illustrations by A. F. JACASSEY. Crown 8vo, 6s.

BOSANQUET, Bernard.—Knowledge and Reality. A Criticism of Mr. F. H. Bradley's "Principles of Logic." Crown 8vo, 9s.

BOUVERIE-PUSEY, S. E. B.—Permanence and Evolution. An Inquiry into the Supposed Mutability of Animal Types. Crown 8vo, 5s.

BOWEN, H. C., M.A.—Studies in English. For the use of Modern Schools. Ninth Thousand. Small crown 8vo, 1s. 6d.

English Grammar for Beginners. Fcap. 8vo, 1s.

Simple English Poems. English Literature for Junior Classes. In four parts. Parts I., II., and III., 6d. each. Part IV., 1s. Complete, 3s.

BRADLEY, F. H.—The Principles of Logic. Demy 8vo, 16s.

BRIDGETT, Rev. T. E.—History of the Holy Eucharist in Great Britain. 2 vols. Demy 8vo, 18s.

BROOKE, Rev. Stopford A.—**The Fight of Faith.** Sermons preached on various occasions. Fifth Edition. Crown 8vo, 7*s.* 6*d.*

The Spirit of the Christian Life. Third Edition. Crown 8vo, 5*s.*

Theology in the English Poets.—Cowper, Coleridge, Wordsworth, and Burns. Sixth Edition. Post 8vo, 5*s.*

Christ in Modern Life. Sixteenth Edition. Crown 8vo, 5*s.*

Sermons. First Series. Thirteenth Edition. Crown 8vo, 5*s.*

Sermons. Second Series. Sixth Edition. Crown 8vo, 5*s.*

BROWN, Horatio F.—**Life on the Lagoons.** With 2 Illustrations and Map. Crown 8vo, 6*s.*

Venetian Studies. Crown 8vo, 7*s.* 6*d.*

BROWN, Rev. J. Baldwin.—**The Higher Life.** Its Reality, Experience, and Destiny. Sixth Edition. Crown 8vo, 5*s.*

Doctrine of Annihilation in the Light of the Gospel of Love. Five Discourses. Fourth Edition. Crown 8vo, 2*s.* 6*d.*

The Christian Policy of Life. A Book for Young Men of Business. Third Edition. Crown 8vo, 3*s.* 6*d.*

BURDETT, Henry C.—**Help in Sickness—Where to Go and What to Do.** Crown 8vo, 1*s.* 6*d.*

Helps to Health. The Habitation—The Nursery—The Schoolroom and—The Person. With a Chapter on Pleasure and Health Resorts. Crown 8vo, 1*s.* 6*d.*

BURKE, Oliver J.—**South Isles of Aran (County Galway).** Crown 8vo, 2*s.* 6*d.*

BURKE, The Late Very Rev. T. N.—**His Life.** By W. J. Fitz-patrick. 2 vols. With Portrait. Demy 8vo, 30*s.*

BURTON, Lady.—**The Inner Life of Syria, Palestine, and the Holy Land.** Post 8vo, 6*s.*

CANDLER, C.—**The Prevention of Consumption.** A Mode of Prevention founded on a New Theory of the Nature of the Tubercle-Bacillus. Demy 8vo, 10*s.* 6*d.*

CAPES, J. M.—**The Church of the Apostles:** an Historical Inquiry. Demy 8vo, 9*s.*

Carlyle and the Open Secret of His Life. By Henry Larkin. Demy 8vo, 14*s.*

CARPENTER, W. B., LL.D., M.D., F.R.S., etc.—**The Principles of Mental Physiology.** With their Applications to the Training and Discipline of the Mind, and the Study of its Morbid Conditions. Illustrated. Sixth Edition. 8vo, 12*s.*

Catholic Dictionary. Containing some Account of the Doctrine, Discipline, Rites, Ceremonies, Councils, and Religious Orders of the Catholic Church. By WILLIAM E. ADDIS and THOMAS ARNOLD, M.A. Third Edition. Demy 8vo, 21*s.*

Century Guild Hobby Horse. Vol. I. Half parchment, 12*s. 6d.*

CHARLES, Rev. R. H.—**Forgiveness,** and other Sermons. Crown 8vo, 4*s. 6d.*

CHEYNE, Canon.—**The Prophecies of Isaiah.** Translated with Critical Notes and Dissertations. 2 vols. Fourth Edition. Demy 8vo, 25*s.*

   Job and Solomon ; or, the Wisdom of the Old Testament. Demy 8vo, 12*s. 6d.*

   The Psalter ; or, The Book of the Praises of Israel. Translated with Commentary. Demy 8vo.

CLAIRAUT. — **Elements of Geometry.** Translated by Dr. KAINES. With 145 Figures. Crown 8vo, 4*s. 6d.*

CLAPPERTON, Jane Hume. — **Scientific Meliorism and the Evolution of Happiness.** Large crown 8vo, 8*s. 6d.*

CLARKE, Rev. Henry James, A.K.C.—**The Fundamental Science.** Demy 8vo, 10*s. 6d.*

CLODD, Edward, F.R.A.S.—**The Childhood of the World : a** Simple Account of Man in Early Times. Eighth Edition. Crown 8vo, 3*s.*
   A Special Edition for Schools. 1*s.*

   The Childhood of Religions. Including a Simple Account of the Birth and Growth of Myths and Legends. Eighth Thousand. Crown 8vo, 5*s.*
   A Special Edition for Schools. 1*s. 6d.*

   Jesus of Nazareth. With a brief sketch of Jewish History to the Time of His Birth. Small crown 8vo, 6*s.*

COGHLAN, J. Cole, D.D.—**The Modern Pharisee and other Sermons.** Edited by the Very Rev. H. H. DICKINSON, D.D., Dean of Chapel Royal, Dublin. New and Cheaper Edition. Crown 8vo, 7*s. 6d.*

COLERIDGE, Sara.—**Memoir and Letters of Sara Coleridge.** Edited by her Daughter. With Index. Cheap Edition. With Portrait. 7*s. 6d.*

COLERIDGE, The Hon. Stephen.—**Demetrius.** Crown 8vo, 5*s.*

CONNELL, A. K.—**Discontent and Danger in India.** Small crown 8vo, 3*s. 6d.*

   The Economic Revolution of India. Crown 8vo, 4*s. 6d.*

COOK, Keningale, LL.D.—**The Fathers of Jesus.** A Study of the Lineage of the Christian Doctrine and Traditions. 2 vols. Demy 8vo, 28*s.*

CORR, the late Rev. T. J., M.A.—**Favilla** ; Tales, Essays, and Poems. Crown 8vo, 5s.

CORY, William.—**A Guide to Modern English History.** Part I. —MDCCCXV.-MDCCCXXX. Demy 8vo, 9s. Part II.— MDCCCXXX.-MDCCCXXXV., 15s.

COTTON, H. J. S.—**New India, or India in Transition.** Third Edition. Crown 8vo, 4s. 6d. ; Cheap Edition, paper covers, 1s.

COUTTS, Francis Burdett Money.—**The Training of the Instinct of Love.** With a Preface by the Rev. EDWARD THRING, M.A. Small crown 8vo, 2s. 6d.

COX, Rev. Sir George W., M.A., Bart.—**The Mythology of the Aryan Nations.** New Edition. Demy 8vo, 16s.

Tales of Ancient Greece. New Edition. Small crown 8vo, 6s.

A Manual of Mythology in the form of Question and Answer. New Edition. Fcap. 8vo, 3s.

An Introduction to the Science of Comparative Mythology and Folk-Lore. Second Edition. Crown 8vo. 7s. 6d.

COX, Rev. Sir G. W., M.A., Bart., and JONES, Eustace Hinton.— **Popular Romances of the Middle Ages.** Third Edition, in 1 vol. Crown 8vo, 6s.

COX, Rev. Samuel, D.D.—**A Commentary on the Book of Job** With a Translation. Second Edition. Demy 8vo, 15s.

Salvator Mundi ; or, 'Is Christ the Saviour of all Men? Tenth Edition. Crown 8vo, 5s.

The Larger Hope. A Sequel to "Salvator Mundi." Second Edition. 16mo, 1s.

The Genesis of Evil, and other Sermons, mainly expository. Third Edition. Crown 8vo, 6s.

Balaam. An Exposition and a Study. Crown 8vo, 5s.

Miracles. An Argument and a Challenge. Crown 8vo, 2s. 6d.

CRAVEN, Mrs.—**A Year's Meditations.** Crown 8vo, 6s.

CRAWFURD, Oswald.—**Portugal, Old and New.** With Illustrations and Maps. New and Cheaper Edition. Crown 8vo, 6s.

CRUISE, Francis Richard, M.D.—**Thomas à Kempis.** Notes of a Visit to the Scenes in which his Life was spent. With Portraits and Illustrations. Demy 8vo, 12s.

CUNNINGHAM, W., B.D —**Politics and Economics :** An Essay on the Nature of the Principles of Political Economy, together with a survey of Recent Legislation. Crown 8vo, 5s.

DANIELL, Clarmont.—**The Gold Treasure of India.** An Inquiry into its Amount, the Cause of its Accumulation, and the Proper Means of using it as Money. Crown 8vo, 5s.

DANIELL, Clarmont.—continued.

Discarded Silver : a Plan for its Use as Money. Small crown 8vo, 2*s.*

DANIEL, Gerard. Mary Stuart : a Sketch and a Defence. Crown 8vo, 5*s.*

DARMESTETER, Arsene.—The Life of Words as the Symbols of Ideas. Crown 8vo, 4*s. 6d.*

DAVIDSON, Rev. Samuel, D.D., LL.D.—Canon of the Bible : Its Formation, History, and Fluctuations. Third and Revised Edition. Small crown 8vo, 5*s.*

The Doctrine of Last Things contained in the New Testament compared with the Notions of the Jews and the Statements of Church Creeds. Small crown 8vo, 3*s. 6d.*

DAWSON, Geo., M.A. Prayers, with a Discourse on Prayer. Edited by his Wife. First Series. Ninth Edition. Crown 8vo, 3*s. 6d.*

Prayers, with a Discourse on Prayer. Edited by GEORGE ST. CLAIR. Second Series. Crown 8vo, 6*s.*

Sermons on Disputed Points and Special Occasions. Edited by his Wife. Fourth Edition. Crown 8vo, 6*s.*

Sermons on Daily Life and Duty. Edited by his Wife. Fourth Edition. Crown 8vo, 6*s.*

The Authentic Gospel, and other Sermons. Edited by GEORGE ST. CLAIR, F.G.S. Third Edition. Crown 8vo, 6*s.*

Biographical Lectures. Edited by GEORGE ST. CLAIR, F.G.S. Third Edition. Large crown 8vo, 7*s. 6d.*

Shakespeare, and other Lectures. Edited by GEORGE ST. CLAIR, F.G.S. Large crown 8vo, 7*s. 6d.*

DE JONCOURT, Madame Marie.—Wholesome Cookery. Fourth Edition. Crown 8vo, cloth, 1*s. 6d*; paper covers, 1*s.*

DENT, H. C.—A Year in Brazil. With Notes on Religion, Meteorology, Natural History, etc. Maps and Illustrations. Demy 8vo, 18*s.*

Doctor Faust. The Old German Puppet Play, turned into English, with Introduction, etc., by T. C. H. HEDDERWICK. Large post 8vo, 7*s. 6d.*

DOWDEN, Edward, LL.D.—Shakspere : a Critical Study of his Mind and Art. Eighth Edition. Post 8vo, 12*s.*

Studies in Literature, 1789–1877. Fourth Edition. Large post 8vo, 6*s.*

Transcripts and Studies. Large post 8vo.

Dulce Domum. Fcap. 8vo, 5*s.*

DU MONCEL, Count.—The Telephone, the Microphone, and the Phonograph. With 74 Illustrations. Third Edition. Small crown 8vo, 5s.

DUNN, H. Percy.—Infant Health. The Physiology and Hygiene of Early Life. Crown 8vo.

DURUY, Victor.—History of Rome and the Roman People. Edited by Prof. MAHAFFY. With nearly 3000 Illustrations. 4to. 6 vols. in 12 parts, 30s. each vol.

Education Library. Edited by Sir PHILIP MAGNUS :—

An Introduction to the History of Educational Theories. By OSCAR BROWNING, M.A. Second Edition. 3s. 6d.

Old Greek Education. By the Rev. Prof. MAHAFFY, M.A. Second Edition. 3s. 6d.

School Management. Including a general view of the work of Education, Organization and Discipline. By JOSEPH LANDON. Sixth Edition. 6s.

EDWARDES, Major-General Sir Herbert B.—Memorials of his Life and Letters. By his Wife. With Portrait and Illustrations. 2 vols. Demy 8vo, 36s.

ELSDALE, Henry.—Studies in Tennyson's Idylls. Crown 8vo, 5s.

Emerson's (Ralph Waldo) Life. By OLIVER WENDELL HOLMES. English Copyright Edition. With Portrait. Crown 8vo, 6s.

"Fan Kwae" at Canton before Treaty Days 1825-1844. By an old Resident. With Frontispiece. Crown 8vo, 5s.

Five o'clock Tea. Containing Receipts for Cakes, Savoury Sandwiches, etc. Fcap. 8vo, cloth, 1s. 6d. ; paper covers, 1s.

FOTHERINGHAM, James.—Studies in the Poetry of Robert Browning. Crown 8vo. 6s.

GARDINER, Samuel R., and J. BASS MULLINGER, M.A.—Introduction to the Study of English History. Second Edition. Large crown 8vo, 9s.

Genesis in Advance of Present Science. A Critical Investigation of Chapters I.-IX. By a Septuagenarian Beneficed Presbyter. Demy 8vo, 10s. 6d.

GEORGE, Henry.—Progress and Poverty : An Inquiry into the Causes of Industrial Depressions, and of Increase of Want with Increase of Wealth. The Remedy. Fifth Library Edition. Post 8vo, 7s. 6d. Cabinet Edition. Crown 8vo, 2s. 6d. Also a Cheap Edition. Limp cloth, 1s. 6d. ; paper covers, 1s.

Protection, or Free Trade. An Examination of the Tariff Question, with especial regard to the Interests of Labour. Second Edition. Crown 8vo, 5s.

GEORGE, Henry.—continued.

> **Social Problems.** Fourth Thousand. Crown 8vo, 5*s.* Cheap Edition, paper covers, 1*s.*

GILBERT, Mrs. — **Autobiography, and other Memorials.** Edited by JOSIAH GILBERT. Fifth Edition. Crown 8vo, 7*s. 6d.*

GLANVILL, Joseph.—**Scepsis Scientifica**; or, Confest Ignorance, the Way to Science; in an Essay of the Vanity of Dogmatizing and Confident Opinion. Edited, with Introductory Essay, by JOHN OWEN. Elzevir 8vo, printed on hand-made paper, 6*s.*

Glossary of Terms and Phrases. Edited by the Rev. H. PERCY SMITH and others. Second and Cheaper Edition. Medium 8vo, 7*s. 6d.*

GLOVER, F., M.A.—**Exempla Latina.** A First Construing Book, with Short Notes, Lexicon, and an Introduction to the Analysis of Sentences. Second Edition. Fcap. 8vo, 2*s.*

GOODENOUGH, Commodore J. G.—**Memoir of,** with Extracts from his Letters and Journals. Edited by his Widow. With Steel Engraved Portrait. Third Edition. Crown 8vo, 5*s.*

GORDON, Major-General C. G.—**His Journals at Kartoum.** Printed from the original MS. With Introduction and Notes by A. EGMONT HAKE. Portrait, 2 Maps, and 30 Illustrations. Two vols., demy 8vo, 21*s.* Also a Cheap Edition in 1 vol., 6*s.*

> **Gordon's (General) Last Journal.** A Facsimile of the last Journal received in England from GENERAL GORDON. Reproduced by Photo-lithography. Imperial 4to, £3 3*s.*

> **Events in his Life.** From the Day of his Birth to the Day of his Death. By Sir H. W. GORDON. With Maps and Illustrations. Second Edition. Demy 8vo, 7*s. 6d.*

GOSSE, Edmund. — **Seventeenth Century Studies.** A Contribution to the History of English Poetry. Demy 8vo, 10*s. 6d.*

GOULD, Rev. S. Baring, M.A.—**Germany, Present and Past.** New and Cheaper Edition. Large crown 8vo, 7*s. 6d.*

> **The Vicar of Morwenstow.** A Life of Robert Stephen Hawker. Crown 8vo, 6*s.*

GOWAN, Major Walter E.—**A. Ivanoff's Russian Grammar.** (16th Edition.) Translated, enlarged, and arranged for use of Students of the Russian Language. Demy 8vo, 6*s.*

GOWER, Lord Ronald. **My Reminiscences.** MINIATURE EDITION, printed on hand-made paper, limp parchment antique, 10*s. 6d.*

> **Bric-à-Brac.** Being some Photoprints taken at Gower Lodge, Windsor. Super royal 8vo.

> **Last Days of Mary Antoinette.** An Historical Sketch. With Portrait and Facsimiles. Fcap. 4to, 10*s. 6d.*

GOWER, Lord Ronald.—continued.

Notes of a Tour from Brindisi to Yokohama, 1883–1884. Fcap. 8vo, 2s. 6d.

GRAHAM, William, M.A.—The Creed of Science, Religious, Moral, and Social. Second Edition, Revised. Crown 8vo, 6s.

The Social Problem, in its Economic, Moral, and Political Aspects. Demy 8vo, 14s.

GREY, Rowland.—In Sunny Switzerland. A Tale of Six Weeks. Second Edition. Small crown 8vo, 5s.

Lindenblumen and other Stories. Small crown 8vo, 5s.

GRIMLEY, Rev. H. N., M.A.—Tremadoc Sermons, chiefly on the Spiritual Body, the Unseen World, and the Divine Humanity. Fourth Edition. Crown 8vo, 6s.

The Temple of Humanity, and other Sermons. Crown 8vo, 6s.

GURNEY, Edmund.—Tertium Quid : chapters on Various Disputed Questions. 2 vols. Crown 8vo, 12s.

HADDON, Caroline.—The Larger Life, Studies in Hinton's Ethics. Crown 8vo, 5s.

HAECKEL, Prof. Ernst.—The History of Creation. Translation revised by Professor E. RAY LANKESTER, M.A., F.R.S. With Coloured Plates and Genealogical Trees of the various groups of both Plants and Animals. 2 vols. Third Edition. Post 8vo, 32s.

The History of the Evolution of Man. With numerous Illustrations. 2 vols. Post 8vo, 32s.

A Visit to Ceylon. Post 8vo, 7s. 6d.

Freedom in Science and Teaching. With a Prefatory Note by T. H. HUXLEY, F.R.S. Crown 8vo, 5s.

Hamilton, Memoirs of Arthur, B.A., of Trinity College, Cambridge. Crown 8vo, 6s.

Handbook of Home Rule, being Articles on the Irish Question by Various Writers. Edited by JAMES BRYCE, M.P. Second Edition. Crown 8vo, 1s. sewed, or 1s. 6d. cloth.

HARRIS, William.—The History of the Radical Party in Parliament. Demy 8vo, 15s.

HAWEIS, Rev. H. R., M.A.—Current Coin. Materialism—The Devil—Crime—Drunkenness—Pauperism—Emotion—Recreation—The Sabbath. Fifth Edition. Crown 8vo, 5s.

Arrows in the Air. Fifth Edition. Crown 8vo, 5s.

Speech in Season. Fifth Edition. Crown 8vo, 5s.

Thoughts for the Times. Fourteenth Edition. Crown 8vo, 5s.

HAWEIS, Rev. H. R., M.A.—continued.

Unsectarian Family Prayers. New Edition. Fcap. 8vo, 1s. 6d.

*HAWTHORNE, Nathaniel.—***Works.** Complete in Twelve Volumes, Large post 8vo, 7s. 6d. each volume.

*HEATH, Francis George.—*Autumnal Leaves. Third and cheaper Edition. Large crown 8vo, 6s.

Sylvan Winter. With 70 Illustrations. Large crown 8vo, 14s.

Hegel's Philosophy of Fine Art. The Introduction, translated by BERNARD BOSANQUET. Crown 8vo, 5s.

*HENNESSY, Sir John Pope.—*Ralegh in Ireland. With his Letters on Irish Affairs and some Contemporary Documents. Large crown 8vo, printed on hand-made paper, parchment, 10s. 6d.

*HENRY, Philip.—*Diaries and Letters of. Edited by MATTHEW HENRY LEE, M.A. Large crown 8vo, 7s. 6d.

*HINTON, J.—*Life and Letters. With an Introduction by Sir W. W. GULL, Bart., and Portrait engraved on Steel by C. H. Jeens. Fifth Edition. Crown 8vo, 8s. 6d.

Philosophy and Religion. Selections from the Manuscripts of the late James Hinton. Edited by CAROLINE HADDON. Second Edition. Crown 8vo, 5s.

The Law Breaker, and The Coming of the Law. Edited by MARGARET HINTON. Crown 8vo, 6s.

The Mystery of Pain. New Edition. Fcap. 8vo, 1s.

Homer's Iliad. Greek text, with a Translation by J. G. CORDERY. 2 vols. Demy 8vo, 24s.

*HOOPER, Mary.—*Little Dinners: How to Serve them with Elegance and Economy. Twentieth Edition. Crown 8vo, 2s. 6d.

Cookery for Invalids, Persons of Delicate Digestion, and Children. Fifth Edition. Crown 8vo, 2s. 6d.

Every-Day Meals. Being Economical and Wholesome Recipes for Breakfast, Luncheon, and Supper. Seventh Edition. Crown 8vo, 2s. 6d.

HOPKINS, Ellice.— Work amongst Working Men. Sixth Edition. Crown 8vo, 3s. 6d.

*HORNADAY, W. T.—*Two Years in a Jungle. With Illustrations. Demy 8vo, 21s.

*HOSPITALIER, E.—*The Modern Applications of Electricity. Translated and Enlarged by JULIUS MAIER, Ph.D. 2 vols. Second Edition, Revised, with many additions and numerous Illustrations. Demy 8vo, 25s.

HOWARD, Robert, M.A.—The Church of England and other Religious Communions. A course of Lectures delivered in the Parish Church of Clapham. Crown 8vo, 7s. 6d.

How to Make a Saint; or, The Process of Canonization in the Church of England. By the PRIG. Fcap 8vo, 3s. 6d.

HUNTER, William C.—Bits of Old China. Small crown 8vo, 6s.

HYNDMAN, H. M.—The Historical Basis of Socialism in England. Large crown 8vo, 8s. 6d.

IDDESLEIGH, Earl of.—The Pleasures, Dangers, and Uses of Desultory Reading. Fcap. 8vo, in Whatman paper cover, 1s.

IM THURN, Everard F.—Among the Indians of Guiana. Being Sketches, chiefly anthropologic, from the Interior of British Guiana. With 53 Illustrations and a Map. Demy 8vo, 18s.

JACCOUD, Prof. S.—The Curability and Treatment of Pulmonary Phthisis. Translated and edited by MONTAGU LUBBOCK, M.D. Demy 8vo, 15s.

Jaunt in a Junk: A Ten Days' Cruise in Indian Seas. Large crown 8vo, 7s. 6d.

JENKINS, E., and RAYMOND, J.—The Architect's Legal Handbook. Third Edition, revised. Crown 8vo, 6s.

JENKINS, Rev. Canon R. C.—Heraldry: English and Foreign. With a Dictionary of Heraldic Terms and 156 Illustrations. Small crown 8vo, 3s. 6d.

The Story of the Caraffa: the Pontificate of Paul IV. Small crown 8vo, 3s. 6d.

JOEL, L.—A Consul's Manual and Shipowner's and Shipmaster's Practical Guide in their Transactions Abroad. With Definitions of Nautical, Mercantile, and Legal Terms; a Glossary of Mercantile Terms in English, French, German, Italian, and Spanish; Tables of the Money, Weights, and Measures of the Principal Commercial Nations and their Equivalents in British Standards; and Forms of Consular and Notarial Acts. Demy 8vo, 12s.

JOHNSTON, H. H., F.Z.S.—The Kilima-njaro Expedition. A Record of Scientific Exploration in Eastern Equatorial Africa, and a General Description of the Natural History, Languages, and Commerce of the Kilima-njaro District. With 6 Maps, and over 80 Illustrations by the Author. Demy 8vo, 21s.

JORDAN, Furneaux, F.R.C.S.—Anatomy and Physiology in Character. Crown 8vo, 5s.

JOYCE, P. W., LL.D., etc.—Old Celtic Romances. Translated from the Gaelic. Crown 8vo, 7s. 6d.

KAUFMANN, Rev. M., B.A.—Socialism : its Nature, its Dangers, and its Remedies considered. Crown 8vo, 7s. 6d.

Utopias ; or, Schemes of Social Improvement, from Sir Thomas More to Karl Marx. Crown 8vo, 5s.

KAY, David, F.R.G.S.—Education and Educators. Crown 8vo. 7s. 6d.

KAY, Joseph.—Free Trade in Land. Edited by his Widow. With Preface by the Right Hon. JOHN BRIGHT, M.P. Seventh Edition. Crown 8vo, 5s.

** Also a cheaper edition, without the Appendix, but with a Review of Recent Changes in the Land Laws of England, by the RIGHT HON. G. OSBORNE MORGAN, Q.C., M.P. Cloth, 1s. 6d. ; paper covers, 1s.

KELKE, W. H. H.—An Epitome of English Grammar for the Use of Students. Adapted to the London Matriculation Course and Similar Examinations: Crown 8vo, 4s. 6d.

KEMPIS, Thomas à.—Of the Imitation of Christ. Parchment Library Edition.—Parchment or cloth, 6s. ; vellum, 7s. 6d. The Red Line Edition, fcap. 8vo, cloth extra, 2s. 6d. The Cabinet Edition, small 8vo, cloth limp, 1s. ; cloth boards, 1s. 6d. The Miniature Edition, cloth limp, 32mo, 1s.

** All the above Editions may be had in various extra bindings.

Notes of a Visit to the Scenes in which his Life was spent. With numerous Illustrations. By F. R. CRUISE, M.D. Demy 8vo, 12s.

KETTLEWELL, Rev. S.—Thomas à Kempis and the Brothers of Common Life. With Portrait. Second Edition. Crown 8vo, 7s. 6d.

KIDD, Joseph, M.D.—The Laws of Therapeutics ; or, the Science and Art of Medicine. Second Edition. Crown 8vo, 6s.

KINGSFORD, Anna, M.D.—The Perfect Way in Diet. A Treatise advocating a Return to the Natural and Ancient Food of our Race. Third Edition. Small crown 8vo, 2s.

KINGSLEY, Charles, M.A.—Letters and Memories of his Life. Edited by his Wife. With two Steel Engraved Portraits, and Vignettes on Wood. Sixteenth Cabinet Edition. 2 vols. Crown 8vo, 12s.

** Also a People's Edition, in one volume. With Portrait. Crown 8vo, 6s.

All Saints' Day, and other Sermons. Edited by the Rev. W. HARRISON. Third Edition. Crown 8vo, 7s. 6d.

True Words for Brave Men. A Book for Soldiers' and Sailors' Libraries. Sixteenth Thousand. Crown 8vo, 2s. 6d.

KNOX, Alexander A.—The New Playground ; or, Wanderings in Algeria. New and Cheaper Edition. Large crown 8vo, 6s.

Kosmos; or, the Hope of the World. 3*s.* 6*d.*

Land Concentration and Irresponsibility of Political Power, as causing the Anomaly of a Widespread State of Want by the Side of the Vast Supplies of Nature. Crown 8vo, 5*s.*

LANDON, Joseph.—**School Management**; Including a General View of the Work of Education, Organization, and Discipline. Sixth Edition. Crown 8vo, 6*s.*

LAURIE, S. S.—**The Rise and Early Constitution of Universities.** With a Survey of Mediæval Education. Crown 8vo, 6*s.*

LEE, Rev. F. G., D.C.L.—**The Other World**; or, Glimpses of the Supernatural. 2 vols. A New Edition. Crown 8vo, 15*s.*

LEFEVRE, 'Right Hon. G. Shaw.—**Peel and O'Connell.** Demy 8vo, 10*s.* 6*d.*

Letters from an Unknown Friend. By the Author of "Charles Lowder." With a Preface by the Rev. W. H. CLEAVER. Fcap. 8vo, 1*s.*

Life of a Prig. By ONE. Third Edition. Fcap. 8vo, 3*s.* 6*d.*

LILLIE, Arthur, M.R.A.S.—**The Popular Life of Buddha.** Containing an Answer to the Hibbert Lectures of 1881. With Illustrations. Crown 8vo, 6*s.*

Buddhism in Christendom; or, Jesus the Essene. With Illustrations. Demy 8vo, 15*s.*

LONGFELLOW, H. Wadsworth.—**Life.** By his Brother, SAMUEL LONGFELLOW. With Portraits and Illustrations. 3 vols. Demy 8vo, 42*s.*

LONSDALE, Margaret.—**Sister Dora**: a Biography. With Portrait. Twenty-ninth Edition. Small crown 8vo, 2*s.* 6*d.*

George Eliot: Thoughts upon her Life, her Books, and Herself. Second Edition. Small crown 8vo, 1*s.* 6*d.*

LOUNSBURY, Thomas R.—**James Fenimore Cooper.** With Portrait. Crown 8vo, 5*s.*

LOWDER, Charles.—**A Biography.** By the Author of "St. Teresa." Twelfth Edition. Crown 8vo. With Portrait. 3*s.* 6*d.*

LÜCKES, Eva C. E.—**Lectures on General Nursing,** delivered to the Probationers of the London Hospital Training School for Nurses. Second Edition. Crown 8vo, 2*s.* 6*d.*

LYALL, William Rowe, D.D.—**Propædeia Prophetica**; or, The Use and Design of the Old Testament Examined. New Edition. With Notices by GEORGE C. PEARSON, M.A., Hon. Canon of Canterbury. Demy 8vo, 10*s.* 6*d.*

LYTTON, Edward Bulwer, Lord.—**Life, Letters and Literary Remains.** By his Son, the EARL OF LYTTON. With Portraits, Illustrations and Facsimiles. Demy 8vo. Vols. I. and II., 32*s.*

MACAULAY, G. C.—Francis Beaumont : A Critical Study. Crown 8vo, 5s.

MACHIAVELLI, Niccolò. — Life and Times. By Prof. VILLARI. Translated by LINDA VILLARI. 4 vols. Large post 8vo, 48s.

Discourses on the First Decade of Titus Livius. Translated from the Italian by NINIAN HILL THOMSON, M.A. Large crown 8vo, 12s.

The Prince. Translated from the Italian by N. H. T. Small crown 8vo, printed on hand-made paper, bevelled boards, 6s.

MACNEILL, J. G. Swift.—How the Union was carried. Crown 8vo, cloth, 1s. 6d. ; paper covers, 1s.

MAGNUS, Lady.—About the Jews since Bible Times. From the Babylonian Exile till the English Exodus. Small crown 8vo, 6s.

MAGUIRE, Thomas.—Lectures on Philosophy. Demy 8vo, 9s.

Many Voices. A volume of Extracts from the Religious Writers of Christendom from the First to the Sixteenth Century. With Biographical Sketches. Crown 8vo, cloth extra, red edges, 6s.

MARKHAM, Capt. Albert Hastings, R.N.—The Great Frozen Sea : A Personal Narrative of the Voyage of the *Alert* during the Arctic Expedition of 1875-6. With 6 full-page Illustrations, 2 Maps, and 27 Woodcuts. Sixth and Cheaper Edition. Crown 8vo, 6s.

MARTINEAU, Gertrude.—Outline Lessons on Morals. Small crown 8vo, 3s. 6d.

MASON, Charlotte M.—Home Education : a Course of Lectures to Ladies. Crown 8vo, 3s. 6d.

Matter and Energy : An Examination of the Fundamental Conceptions of Physical Force. By B. L. L. Small crown 8vo, 2s.

MAUDSLEY, H., M.D.—Body and Will. Being an Essay concerning Will, in its Metaphysical, Physiological, and Pathological Aspects. 8vo, 12s.

Natural Causes and Supernatural Seemings. Second Edition. Crown 8vo, 6s.

McGRATH, Terence.—Pictures from Ireland. New and Cheaper Edition. Crown 8vo, 2s.

MEREDITH, M.A.—Theotokos, the Example for Woman. Dedicated, by permission, to Lady Agnes Wood. Revised by the Venerable Archdeacon DENISON. 32mo, limp cloth, 1s. 6d.

MILLER, Edward.—The History and Doctrines of Irvingism ; or, The so-called Catholic and Apostolic Church. 2 vols. Large post 8vo, 15s.

The Church in Relation to the State. Large crown 8vo, 4s.

MILLS, Herbert.—**Poverty and the State ;** or, Work for the Unemployed. An Inquiry into the Causes and Extent of Enforced Idleness, with a Statement of a Remedy. Crown 8vo, 6s.

MITCHELL, Lucy M.—**A History of Ancient Sculpture.** With numerous Illustrations, including 6 Plates in Phototype. Super-royal 8vo, 42s.

MOCKLER, E.—**A Grammar of the Baloochee Language,** as it is spoken in Makran (Ancient Gedrosia), in the Persia-Arabic and Roman characters. Fcap. 8vo, 5s.

MOHL, Julius and Mary.—**Letters and Recollections of.** By M. C. M. SIMPSON. With Portraits and Two Illustrations. Demy 8vo, 15s.

MOLESWORTH, Rev. W. Nassau, M.A.—**History of the Church of England from 1660.** Large crown 8vo, 7s. 6d.

MORELL, J. R.—**Euclid Simplified in Method and Language.** Being a Manual of Geometry. Compiled from the most important French Works, approved by the University of Paris and the Minister of Public Instruction. Fcap. 8vo, 2s. 6d.

MORGAN, C. Lloyd.—**The Springs of Conduct.** An Essay in Evolution. Large crown 8vo, cloth, 7s. 6d.

MORISON, J. Cotter.—**The Service of Man :** an Essay towards the Religion of the Future. Second Edition. Demy 8vo, 10s. 6d.

MORSE, E. S., Ph.D.—**First Book of Zoology.** With numerous Illustrations. New and Cheaper Edition. Crown 8vo, 2s. 6d.

My Lawyer : A Concise Abridgment of the Laws of England. By a Barrister-at-Law. Crown 8vo, 6s. 6d.

NELSON, J. H., M.A.—**A Prospectus of the Scientific Study of the Hindû Law.** Demy 8vo, 9s.

  Indian Usage and Judge-made Law in Madras. Demy 8vo, 12s.

NEWMAN, Cardinal.—**Characteristics from the Writings of.** Being Selections from his various Works. Arranged with the Author's personal Approval. Seventh Edition. With Portrait. Crown 8vo, 6s.

  ** A Portrait of Cardinal Newman, mounted for framing, can be had, 2s. 6d.

NEWMAN, Francis William.—**Essays on Diet.** Small crown 8vo, cloth limp, 2s.

New Social Teachings. By POLITICUS. Small crown 8vo, 5s.

NICOLS, Arthur, F.G.S., F.R.G.S.—**Chapters from the Physical History of the Earth :** an Introduction to Geology and Palæontology. With numerous Illustrations. Crown 8vo, 5s.

NIHILL, Rev. H. D.—**The Sisters of St. Mary at the Cross :** Sisters of the Poor and their Work. Crown 8vo, 2s. 6d.

NOEL, The Hon. Roden.—Essays on Poetry and Poets. Demy 8vo, 12s.

NOPS, Marianne.—Class Lessons on Euclid. Part I. containing the First Two Books of the Elements. Crown 8vo, 2s. 6d.

Nuces: EXERCISES ON THE SYNTAX OF THE PUBLIC SCHOOL LATIN PRIMER. New Edition in Three Parts. Crown 8vo, each 1s.
. The Three Parts can also be had bound together, 3s.

OATES, Frank, F.R.G.S.—Matabele Land and the Victoria Falls. A Naturalist's Wanderings in the Interior of South Africa. Edited by C. G. OATES, B.A. With numerous Illustrations and 4 Maps. Demy 8vo, 21s.

O'BRIEN, R. Barry.—Irish Wrongs and English Remedies, with other Essays. Crown 8vo, 5s.

OGLE, Anna C.—A Lost Love. Small crown 8vo, 2s. 6d.

O'MEARA, Kathleen.—Henri Perreyve and his Counsels to the Sick. Small crown 8vo, 5s.

One and a Half in Norway. A Chronicle of Small Beer. By Either and Both. Small crown 8vo, 3s. 6d.

O'NEIL, the late Rev. Lord.—Sermons. With Memoir and Portrait. Crown 8vo, 6s.

　　Essays and Addresses. Crown 8vo, 5s.

OTTLEY, H. Bickersteth.—The Great Dilemma. Christ His Own Witness or His Own Accuser. Six Lectures. Second Edition. Crown 8vo, 3s. 6d.

Our Public Schools—Eton, Harrow, Winchester, Rugby, Westminster, Marlborough, The Charterhouse. Crown 8vo, 6s.

PADGHAM, Richard.—In the Midst of Life we are in Death. Crown 8vo, 5s.

PALMER, the late William.—Notes of a Visit to Russia in 1840-1841. Selected and arranged by JOHN H. CARDINAL NEWMAN, with Portrait. Crown 8vo, 8s. 6d.

　　Early Christian Symbolism. A Series of Compositions from Fresco Paintings, Glasses, and Sculptured Sarcophagi. Edited by the Rev. Provost NORTHCOTE, D.D., and the Rev. Canon BROWNLOW, M.A. With Coloured Plates, folio, 42s., or with Plain Plates, folio, 25s.

Parchment Library. Choicely Printed on hand-made paper, limp parchment antique or cloth, 6s. ; vellum, 7s. 6d. each volume.

　　The Poetical Works of John Milton. 2 vols.

　　Chaucer's Canterbury Tales. Edited by A. W. POLLARD. 2 vols.

Parchment Library—*continued.*

Letters and Journals of Jonathan Swift. Selected and edited, with a Commentary and Notes, by STANLEY LANE POOLE.

De Quincey's Confessions of an English Opium Eater. Reprinted from the First Edition. Edited by RICHARD GARNETT.

The Gospel according to Matthew, Mark, and Luke.

Selections from the Prose Writings of Jonathan Swift. With a Preface and Notes by STANLEY LANE-POOLE and Portrait.

English Sacred Lyrics.

Sir Joshua Reynolds's Discourses. Edited by EDMUND GOSSE.

Selections from Milton's Prose Writings. Edited by ERNEST MYERS.

The Book of Psalms. Translated by the Rev. Canon T. K. CHEYNE, M.A., D.D.

The Vicar of Wakefield. With Preface and Notes by AUSTIN DOBSON.

English Comic Dramatists. Edited by OSWALD CRAWFURD.

English Lyrics.

The Sonnets of John Milton. Edited by MARK PATTISON. With Portrait after Vertue.

French Lyrics. Selected and Annotated by GEORGE SAINTS-BURY. With a Miniature Frontispiece designed and etched by H. G. Glindoni.

Fables by Mr. John Gay. With Memoir by AUSTIN DOBSON, and an Etched Portrait from an unfinished Oil Sketch by Sir Godfrey Kneller.

Select Letters of Percy Bysshe Shelley. Edited, with an Introduction, by RICHARD GARNETT.

The Christian Year. Thoughts in Verse for the Sundays and Holy Days throughout the Year. With Miniature Portrait of the Rev. J. Keble, after a Drawing by G. Richmond, R.A.

Shakspere's Works. Complete in Twelve Volumes.

Eighteenth Century Essays. Selected and Edited by AUSTIN DOBSON. With a Miniature Frontispiece by R. Caldecott.

Q. Horati Flacci Opera. Edited by F. A. CORNISH, Assistant Master at Eton. With a Frontispiece after a design by L. Alma Tadema, etched by Leopold Lowenstam.

Edgar Allan Poe's Poems. With an Essay on his Poetry by ANDREW LANG, and a Frontispiece by Linley Sambourne.

Parchment Library—*continued.*

> **Shakspere's Sonnets.** Edited by EDWARD DOWDEN. With a Frontispiece etched by Leopold Lowenstam, after the Death Mask.
>
> **English Odes.** Selected by EDMUND GOSSE. With Frontispiece on India paper by Hamo Thornycroft, A.R.A.
>
> **Of the Imitation of Christ.** By THOMAS À KEMPIS. A revised Translation. With Frontispiece on India paper, from a Design by W. B. Richmond.
>
> **Poems:** Selected from PERCY BYSSHE SHELLEY. Dedicated to Lady Shelley. With a Preface by RICHARD GARNETT and a Miniature Frontispiece.

PARSLOE, Joseph.—**Our Railways.** Sketches, Historical and Descriptive. With Practical Information as to Fares and Rates, etc., and a Chapter on Railway Reform. Crown 8vo, 6s.

PASCAL, Blaise.—**The Thoughts of.** Translated from the Text of Auguste Molinier, by C. KEGAN PAUL. Large crown 8vo, with Frontispiece, printed on hand-made paper, parchment antique, or cloth, 12s. ; vellum, 15s.

PAUL, Alexander.—**Short Parliaments.** A History of the National Demand for frequent General Elections. Small crown 8vo, 3s. 6d.

PAUL, C. Kegan.—**Biographical Sketches.** Printed on hand-made paper, bound in buckram. Second Edition. Crown 8vo, 7s. 6d.

PEARSON, Rev. S.—**Week-day Living.** A Book for Young Men and Women. Second Edition. Crown 8vo, 5s.

PENRICE, Major J.—**Arabic and English Dictionary of the Koran.** 4to, 21s.

PESCHEL, Dr. Oscar.—**The Races of Man and their Geographical Distribution.** Second Edition. Large crown 8vo, 9s.

PIDGEON, D.—**An Engineer's Holiday ;** or, Notes of a Round Trip from Long. 0° to 0°. New and Cheaper Edition. Large crown 8vo, 7s. 6d.

> **Old World Questions and New World Answers.** Second Edition. Large crown 8vo, 7s. 6d.

Plain Thoughts for Men. Eight Lectures delivered at Forester's Hall, Clerkenwell, during the London Mission, 1884. Crown 8vo, cloth, 1s. 6d ; paper covers, 1s.

PRICE, Prof. Bonamy. — **Chapters on Practical Political Economy.** Being the Substance of Lectures delivered before the University of Oxford. New and Cheaper Edition. Crown 8vo, 5s.

Prig's Bede : the Venerable Bede, Expurgated, Expounded, and Exposed. By The Prig. Second Edition. Fcap. 8vo, 3s. 6d.

Pulpit Commentary, The. (*Old Testament Series.*) Edited by the Rev. J. S. EXELL, M.A., and the Very Rev. Dean H. D. M. SPENCE, M.A., D.D.

Genesis. By the Rev. T. WHITELAW, D.D. With Homilies by the Very Rev. J. F. MONTGOMERY, D.D., Rev. Prof. R. A. REDFORD, M.A., LL.B., Rev. F. HASTINGS, Rev. W. ROBERTS, M.A. An Introduction to the Study of the Old Testament by the Venerable Archdeacon FARRAR, D.D., F.R.S.; and Introductions to the Pentateuch by the Right Rev. H. COTTERILL, D.D., and Rev. T. WHITELAW, M.A. Eighth Edition. 1 vol., 15*s.*

Exodus. By the Rev. Canon RAWLINSON. With Homilies by Rev. J. ORR, Rev. D. YOUNG, B.A., Rev. C. A. GOODHART, Rev. J. URQUHART, and the Rev. H. T. ROBJOHNS. Fourth Edition. 2 vols., 18*s.*

Leviticus. By the Rev. Prebendary MEYRICK, M.A. With Introductions by the Rev. R. COLLINS, Rev. Professor A. CAVE, and Homilies by Rev. Prof. REDFORD, LL.B., Rev. J. A. MACDONALD, Rev. W. CLARKSON, B.A., Rev. S. R. ALDRIDGE, LL.B., and Rev. MCCHEYNE EDGAR. Fourth Edition. 15*s.*

Numbers. By the Rev. R. WINTERBOTHAM, LL.B. With Homilies by the Rev. Professor W. BINNIE, D.D., Rev. E. S. PROUT, M.A., Rev. D. YOUNG, Rev. J. WAITE, and an Introduction by the Rev. THOMAS WHITELAW, M.A. Fifth Edition. 15*s.*

Deuteronomy. By the Rev. W. L. ALEXANDER, D.D. With Homilies by Rev. C. CLEMANCE, D.D., Rev. J. ORR, B.D., Rev. R. M. EDGAR, M.A., Rev. D. DAVIES, M.A. Fourth edition. 15*s.*

Joshua. By Rev. J. J. LIAS, M.A. With Homilies by Rev. S. R. ALDRIDGE, LL.B., Rev. R. GLOVER, REV. E. DE PRESSENSÉ, D.D., Rev. J. WAITE, B.A., Rev. W. F. ADENEY, M.A.; and an Introduction by the Rev. A. PLUMMER, M.A. Fifth Edition. 12*s.* 6*d.*

Judges and Ruth. By the Bishop of BATH and WELLS, and Rev. J. MORISON, D.D. With Homilies by Rev. A. F. MUIR, M.A., Rev. W. F. ADENEY, M.A., Rev. W. M. STATHAM, and Rev. Professor J. THOMSON, M.A. Fifth Edition. 10*s.* 6*d.*

1 Samuel. By the Very Rev. R. P. SMITH, D.D. With Homilies by Rev. DONALD FRASER, D.D., Rev. Prof. CHAPMAN, and Rev. B. DALE. Sixth Edition. 15*s.*

1 Kings. By the Rev. JOSEPH HAMMOND, LL.B. With Homilies by the Rev. E. DE PRESSENSÉ, D.D., Rev. J. WAITE, B.A., Rev. A. ROWLAND, LL.B., Rev. J. A. MACDONALD, and Rev. J. URQUHART. Fifth Edition. 15*s.*

Pulpit Commentary, The—*continued.*

1 Chronicles. By the Rev. Prof. P. C. BARKER, M.A., LL.B. With Homilies by Rev. Prof. J. R. THOMSON, M.A., Rev. R. TUCK, B.A., Rev. W. CLARKSON, B.A., Rev. F. WHITFIELD, M.A., and Rev. RICHARD GLOVER. 15*s.*

Ezra, Nehemiah, and Esther. By Rev. Canon G. RAWLINSON, M.A. With Homilies by Rev. Prof. J. R. THOMSON, M.A., Rev. Prof. R. A. REDFORD, LL.B., M.A., Rev. W. S. LEWIS, M.A., Rev. J. A. MACDONALD, Rev. A. MACKENNAL, B.A., Rev. W. CLARKSON, B.A., Rev. F. HASTINGS, Rev. W. DINWIDDIE, LL.B., Rev. Prof. ROWLANDS, B.A., Rev. G. WOOD, B.A., Rev. Prof. P. C. BARKER, M.A., LL.B., and the Rev. J. S. EXELL, M.A. Sixth Edition. 1 vol., 12*s.* 6*d.*

Isaiah. By the Rev. Canon G. RAWLINSON, M.A. With Homilies by Rev. Prof. E. JOHNSON, M.A., Rev. W. CLARKSON, B.A., Rev. W. M. STATHAM, and Rev. R. TUCK, B.A. Second Edition. 2 vols., 15*s.* each.

Jeremiah. (Vol. I.) By the Rev. Canon T. K. CHEYNE, M.A., D.D. With Homilies by the Rev. W. F. ADENEY, M.A., Rev. A. F. MUIR, M.A., Rev. S. CONWAY, B.A., Rev. J. WAITE, B.A., and Rev. D. YOUNG, B.A. Third Edition. 15*s.*

Jeremiah (Vol. II.) and Lamentations. By Rev. T. K. CHEYNE, M.A. With Homilies by Rev. Prof. J. R. THOMSON, M.A., Rev. W. F. ADENEY, M.A., Rev. A. F. MUIR, M.A., Rev. S. CONWAY, B.A., Rev. D. YOUNG, B.A. 15*s.*

Hosea and Joel. By the Rev. Prof. J. J. GIVEN, Ph.D., D.D. With Homilies by the Rev. Prof. J. R. THOMSON, M.A., Rev. A. ROWLAND, B.A., LL.B., Rev. C. JERDAN, M.A., LL.B., Rev. J. ORR, M.A., B.D., and Rev. D. THOMAS, D.D. 15*s.*

Pulpit Commentary, The. (*New Testament Series.*)
St. Mark. By Very Rev. E. BICKERSTETH, D.D., Dean of Lichfield. With Homilies by Rev. Prof. THOMSON, M.A., Rev. Prof. J. J. GIVEN, Ph.D., D.D., Rev. Prof. JOHNSON, M.A., Rev. A. ROWLAND, B.A., LL.B., Rev. A. MUIR, and Rev. R. GREEN. Fifth Edition. 2 vols., 21*s.*

The Acts of the Apostles. By the Bishop of BATH and WELLS. With Homilies by Rev. Prof. P. C. BARKER, M.A., LL.B., Rev. Prof. E. JOHNSON, M.A., Rev. Prof. R. A. REDFORD, LL.B., Rev. R. TUCK, B.A., Rev. W. CLARKSON, B.A. Third Edition. 2 vols., 21*s.*

1 Corinthians. By the Ven. Archdeacon FARRAR, D.D. With Homilies by Rev. Ex-Chancellor LIPSCOMB, LL.D., Rev. DAVID THOMAS, D.D., Rev. D. FRASER, D.D., Rev. Prof. J. R. THOMSON, M.A., Rev. J. WAITE, B.A., Rev. R. TUCK, B.A., Rev. E. HURNDALL, M.A., and Rev. H. BREMNER, B.D. Third Edition. 15*s.*

Pulpit Commentary, The—*continued.*

2 Corinthians and Galatians. By the Ven. Archdeacon FARRAR, D.D., and Rev. Prebendary E. HUXTABLE. With Homilies by Rev. Ex-Chancellor LIPSCOMB, LL.D., Rev. DAVID THOMAS, D.D., Rev. DONALD FRASER, D.D., Rev. R. TUCK, B.A., Rev. E. HURNDALL, M.A., Rev. Prof. J. R. THOMSON, M.A., Rev. R. FINLAYSON, B.A., Rev. W. F. ADENEY, M.A., Rev. R. M. EDGAR, M.A., and Rev. T. CROSKERY, D.D. 21*s.*

Ephesians, Philippians, and Colossians. By the Rev. Prof. W. G. BLAIKIE, D.D., Rev. B. C. CAFFIN, M.A., and Rev. G. G. FINDLAY, B.A. With Homilies by Rev. D. THOMAS, D.D., Rev. R. M. EDGAR, M.A., Rev. R. FINLAYSON, B.A., Rev. W. F. ADENEY, M.A., Rev. Prof. T. CROSKERY, D.D., Rev. E. S. PROUT, M.A., Rev. Canon VERNON HUTTON, and Rev. U. R. THOMAS, D.D. Second Edition. 21*s.*

Thessalonians, Timothy, Titus, and Philemon. By the Bishop of Bath and Wells, Rev. Dr. GLOAG and Rev. Dr. EALES. With Homilies by the Rev. B. C. CAFFIN, M.A., Rev. R. FINLAYSON, B.A., Rev. Prof. T. CROSKERY, D.D., Rev. W. F. ADENEY, M.A., Rev. W. M. STATHAM, and Rev. D. THOMAS, D.D. 15*s.*

Hebrews and James. By the Rev. J. BARMBY, D.D., and Rev Prebendary E. C. S. GIBSON, M.A. With Homiletics by the Rev. C. JERDAN, M.A., LL.B., and Rev. Prebendary E. C. S. GIBSON. And Homilies by the Rev. W. JONES, Rev. C. NEW, Rev. D. YOUNG, B.A., Rev. J. S. BRIGHT, Rev. T. F. LOCKYER, B.A., and Rev. C. JERDAN, M.A., LL.B. Second Edition. 15*s.*

PUSEY, Dr.—Sermons for the Church's Seasons from Advent to Trinity. Selected from the Published Sermons of the late EDWARD BOUVERIE PUSEY, D.D. Crown 8vo, 5*s.*

RANKE, Leopold von.—Universal History. The oldest Historical Group of Nations and the Greeks. Edited by G. W. PROTHERO. Demy 8vo, 16*s.*

RENDELL, J. M.—Concise Handbook of the Island of Madeira. With Plan of Funchal and Map of the Island. Fcap. 8vo, 1*s.* 6*d.*

REVELL, W. F.—Ethical Forecasts. Crown 8vo.

REYNOLDS, Rev. J. W.—The Supernatural in Nature. A Verification by Free Use of Science. Third Edition, Revised and Enlarged. Demy 8vo, 14*s.* .

The Mystery of Miracles. Third and Enlarged Edition. Crown 8vo, 6*s.*

The Mystery of the Universe our Common Faith. Demy 8vo, 14*s.*

The World to Come: Immortality a Physical Fact. Crown 8vo, 6*s.*

RIBOT, Prof. Th.—Heredity: A Psychological Study of its Phenomena, its Laws, its Causes, and its Consequences. Second Edition. Large crown 8vo, 9*s.*

ROBERTSON, The late Rev. F. W., M.A.—Life and Letters of. Edited by the Rev. STOPFORD BROOKE, M.A.
    I. Two vols., uniform with the Sermons. With Steel Portrait. Crown 8vo, 7*s.* 6*d.*
    II. Library Edition, in Demy 8vo, with Portrait. 12*s.*
    III. A Popular Edition, in 1 vol. Crown 8vo, 6*s.*

ROBERTSON, The late Rev. F. W., M.A.—continued.

Sermons. Four Series. Small crown 8vo, 3*s.* 6*d.* each.

The Human Race, and other Sermons. Preached at Cheltenham, Oxford, and Brighton. New and Cheaper Edition. Small crown 8vo, 3*s.* 6*d.*

Notes on Genesis. New and Cheaper Edition. Small crown 8vo, 3*s.* 6*d.*

Expository Lectures on St. Paul's Epistles to the Corinthians. A New Edition. Small crown 8vo, 5*s.*

Lectures and Addresses, with other Literary Remains. A New Edition. Small crown 8vo, 5*s.*

An Analysis of Tennyson's "In Memoriam." (Dedicated by Permission to the Poet-Laureate.) Fcap. 8vo, 2*s.*

The Education of the Human Race. Translated from the German of GOTTHOLD EPHRAIM LESSING. Fcap. 8vo, 2*s.* 6*d.*

The above Works can also be had, bound in half morocco.

  ** A Portrait of the late Rev. F. W. Robertson, mounted for framing, can be had, 2*s.* 6*d.*

ROMANES, G. J.—Mental Evolution in Animals. With a Posthumous Essay on Instinct by CHARLES DARWIN, F.R.S. Demy 8vo, 12*s.*

ROOSEVELT, Theodore. Hunting Trips of a Ranchman. Sketches of Sport on the Northern Cattle Plains. With 26 Illustrations. Royal 8vo, 18*s.*

ROSMINI SERBATI, Antonio.—Life. By the REV. W. LOCKHART. Second Edition. 2 vols. With Portraits. Crown 8vo, 12*s.*

Rosmini's Origin of Ideas. Translated from the Fifth Italian Edition of the Nuovo Saggio *Sull' origine delle idee.* 3 vols. Demy 8vo, cloth, 10*s.* 6*d.* each.

Rosmini's Psychology. 3 vols. Demy 8vo [Vols. I. and II. now ready], 10*s.* 6*d.* each.

ROSS, Janet.—Italian Sketches. With 14 full-page Illustrations. Crown 8vo, 7*s.* 6*d.*

RULE, Martin, M.A.—The Life and Times of St. Anselm, Archbishop of Canterbury and Primate of the Britains. 2 vols. Demy 8vo, 32*s.*

SAMUELL, Richard.—Seven, the Sacred Number : Its use in Scripture and its Application to Biblical Criticism. Crown 8vo, 10s. 6d.

SAYCE, Rev. Archibald Henry.—Introduction to the Science of Language. 2 vols. Second Edition. Large post 8vo, 21s.

SCOONES, W. Baptiste.—Four Centuries of English Letters : A Selection of 350 Letters by 150 Writers, from the Period of the Paston Letters to the Present Time. Third Edition. Large crown 8vo, 6s.

SÉE, Prof. Germain.—Bacillary Phthisis of the Lungs. Translated and edited for English Practitioners by WILLIAM HENRY WEDDELL, M.R.C.S. Demy 8vo, 10s. 6d.

Shakspere's Works. The Avon Edition, 12 vols., fcap. 8vo, cloth, 18s. ; in cloth box, 21s. ; bound in 6 vols., cloth, 15s.

Shakspere's Works, an Index to. By EVANGELINE O'CONNOR. Crown 8vo, 5s.

SHELLEY, Percy Bysshe.—Life. By EDWARD DOWDEN, LL.D. 2 vols. With Portraits. Demy 8vo, 36s.

SHILLITO, Rev. Joseph.—Womanhood : its Duties, Temptations, and Privileges. A Book for Young Women. Third Edition. Crown 8vo, 3s. 6d.

Shooting, Practical Hints. Being a Treatise on the Shot Gun and its Management. By "20 Bore." With 55 Illustrations. Demy 8vo, 12s.

Sister Augustine, Superior of the Sisters of Charity at the St. Johannis Hospital at Bonn. Authorized Translation by HANS THARAU, from the German "Memorials of AMALIE VON LASAULX." Cheap Edition. Large crown 8vo, 4s. 6d.

SKINNER, James.—A Memoir. By the Author of "Charles Lowder." With a Preface by the Rev. Canon CARTER, and Portrait. Large crown, 7s. 6d.

*** Also a cheap Edition. With Portrait. Fourth Edition. Crown 8vo, 3s. 6d.

SMEATON, D. Mackenzie. — The Loyal Karens of Burma. Crown 8vo, 4s. 6d.

SMITH, Edward, M.D., LL.B., F.R.S.—Tubercular Consumption in its Early and Remediable Stages. Second Edition. Crown 8vo, 6s.

SMITH, Sir W. Cusack, Bart.—Our War Ships. A Naval Essay. Crown 8vo, 5s.

Spanish Mystics. By the Editor of "Many Voices." Crown 8vo, 5s.

Specimens of English Prose Style from Malory to Macaulay. Selected and Annotated, with an Introductory Essay, by GEORGE SAINTSBURY. Large crown 8vo, printed on handmade paper, parchment antique or cloth, 12s. ; vellum, 15s.

SPEDDING, James.—Reviews and Discussions, Literary, Political, and Historical not relating to Bacon. Demy 8vo, 12s. 6d.

Evenings with a Reviewer; or, Macaulay and Bacon. With a Prefatory Notice by G. S. VENABLES, Q.C. 2 vols. Demy 8vo, 18s.

Stray Papers on Education, and Scenes from School Life. By B. H. Second Edition. Small crown 8vo, 3s. 6d.

STREATFEILD, Rev. G. S., M.A.—Lincolnshire and the Danes. Large crown 8vo, 7s. 6d.

STRECKER-WISLICENUS.—Organic Chemistry. Translated and Edited, with Extensive Additions, by W. R. HODGKINSON, Ph.D., and A. J. GREENAWAY, F.I.C. Second and cheaper Edition. Demy 8vo, 12s. 6d.

Suakin, 1885; being a Sketch of the Campaign of this year. By an Officer who was there. Second Edition. Crown 8vo, 2s. 6d.

SULLY, James, M.A.—Pessimism : a History and a Criticism. Second Edition. Demy 8vo, 14s.

Sunshine and Sea. A Yachting Visit to the Channel Islands and Coast of Brittany. With Frontispiece from a Photograph and 24 Illustrations. Crown 8vo, 6s.

SWEDENBORG, Eman.—De Cultu et Amore Dei ubi Agitur de Telluris ortu, Paradiso et Vivario, tum de Primogeniti Seu Adami Nativitate Infantia, et Amore. Crown 8vo, 6s.

On the Worship and Love of God. Treating of the Birth of the Earth, Paradise, and the Abode of Living Creatures. Translated from the original Latin. Crown 8vo, 7s. 6d.

Prodromus Philosophiæ Ratiocinantis de Infinito, et Causa Finali Creationis : deque Mechanismo Operationis Animæ et Corporis. Edidit THOMAS MURRAY GORMAN, M.A. Crown 8vo, 7s. 6d.

TACITUS.—The Agricola. A Translation. Small crown 8vo, 2s. 6d.

TARRING, C. J.—A Practical Elementary Turkish Grammar. Crown 8vo, 6s.

TAYLOR, Rev. Isaac.—The Alphabet. An Account of the Origin and Development of Letters. With numerous Tables and Facsimiles. 2 vols. Demy 8vo, 36s.

TAYLOR, Jeremy.—The Marriage Ring. With Preface, Notes, and Appendices. Edited by FRANCIS BURDETT MONEY COUTTS. Small crown 8vo, 2s. 6d.

TAYLOR, Sedley. — Profit Sharing between Capital and Labour. To which is added a Memorandum on the Industrial Partnership at the Whitwood Collieries, by ARCHIBALD and HENRY BRIGGS, with remarks by SEDLEY TAYLOR. Crown 8vo, 2s. 6d.

THOM, J. Hamilton.—Laws of Life after the Mind of Christ. Two Series. Crown 8vo, 7s. 6d. each.

THOMPSON, Sir H.—Diet in Relation to Age and Activity. Fcap. 8vo, cloth, 1s. 6d. ; paper covers, 1s.

TIDMAN, Paul F.—Money and Labour. 1s. 6d.

TIPPLE, Rev. S. A.—Sunday Mornings at Norwood. Prayers and Sermons. Crown 8vo, 6s.

TODHUNTER, Dr. J.—A Study of Shelley. Crown 8vo, 7s.

TOLSTOI, Count Leo.—Christ's Christianity. Translated from the Russian. Large crown 8vo, 7s. 6d. :

TRANT, William.—Trade Unions : Their Origin, Objects, and Efficacy. Small crown 8vo, 1s. 6d. ; paper covers, 1s.

TRENCH, The late R. C., Archbishop.—Notes on the Parables of Our Lord. Fourteenth Edition. 8vo, 12s. Cheap Edition, 7s. 6d.

Notes on the Miracles of Our Lord. Twelfth Edition. 8vo, 12s. Cheap Edition, 7s. 6d.

Studies in the Gospels. Fifth Edition, Revised. 8vo, 10s. 6d.

Brief Thoughts and Meditations on Some Passages in Holy Scripture. Third Edition. Crown 8vo, 3s. 6d.

Synonyms of the New Testament. Tenth Edition, Enlarged. 8vo, 12s.

Sermons New and Old. Crown 8vo, 6s.

On the Authorized Version of the New Testament. Second Edition. 8vo, 7s.

Commentary on the Epistles to the Seven Churches in Asia. Fourth Edition, Revised. 8vo, 8s. 6d.

The Sermon on the Mount. An Exposition drawn from the Writings of St. Augustine, with an Essay on his Merits as an Interpreter of Holy Scripture. Fourth Edition, Enlarged. 8vo, 10s. 6d.

Shipwrecks of Faith. Three Sermons preached before the University of Cambridge in May, 1867. Fcap. 8vo, 2s. 6d.

Lectures on Mediæval Church History. Being the Substance of Lectures delivered at Queen's College, London. Second Edition. 8vo, 12s.

English, Past and Present. Thirteenth Edition, Revised and Improved. Fcap. 8vo, 5s.

On the Study of Words. Nineteenth Edition, Revised. Fcap. 8vo, 5s.

TRENCH, The late R. C., Archbishop.—continued.

Select Glossary of English Words Used Formerly in Senses Different from the Present. Sixth Edition, Revised and Enlarged. Fcap. 8vo, 5s.

Proverbs and Their Lessons. Seventh Edition, Enlarged. Fcap. 8vo, 4s.

Poems. Collected and Arranged anew. Ninth Edition. Fcap. 8vo, 7s. 6d.

Poems. Library Edition. 2 vols. Small crown 8vo, 10s.

Sacred Latin Poetry. Chiefly Lyrical, Selected and Arranged for Use. Third Edition, Corrected and Improved. Fcap. 8vo, 7s.

A Household Book of English Poetry. Selected and Arranged, with Notes. Fourth Edition, Revised. Extra fcap. 8vo, 5s. 6d.

An Essay on the Life and Genius of Calderon. With Translations from his "Life's a Dream" and "Great Theatre of the World." Second Edition, Revised and Improved. Extra fcap. 8vo, 5s. 6d.

Gustavus Adolphus in Germany, and other Lectures on the Thirty Years' War. Third Edition, Enlarged. Fcap. 8vo, 4s.

Plutarch : his Life, his Lives, and his Morals. Second Edition, Enlarged. Fcap. 8vo, 3s. 6d.

Remains of the late Mrs. Richard Trench. Being Selections from her Journals, Letters, and other Papers. New and Cheaper Issue. With Portrait. 8vo, 6s.

*TUKE, Daniel Hack, M.D., F.R.C.P.—*Chapters in the History of the Insane in the British Isles. With Four Illustrations. Large crown 8vo, 12s.

*TWINING, Louisa.—*Workhouse Visiting and Management during Twenty-Five Years. Small crown 8vo, 2s.

*VAUGHAN, H. Halford.—*New Readings and Renderings of Shakespeare's Tragedies. 3 vols. Demy 8vo, 12s. 6d. each.

*VICARY, J. Fulford.—*Saga Time. With Illustrations. Crown 8vo, 7s. 6d.

*VOGT, Lieut.-Col. Hermann.—*The Egyptian War of 1882. A translation. With Map and Plans. Large crown 8vo, 6s.

*VOLCKXSOM, E. W. v.—*Catechism of Elementary Modern Chemistry. Small crown 8vo, 3s.

*WALPOLE, Chas. George.—*A Short History of Ireland from the Earliest Times to the Union with Great Britain. With 5 Maps and Appendices. Third Edition. Crown 8vo, 6s.

WARD, Wilfrid.—The Wish to Believe. A Discussion Concerning the Temper of Mind, in which a reasonable Man should undertake Religious Inquiry. Small crown 8vo, 5*s.* '

WARD, William George, Ph.D.—Essays on the Philosophy of Theism. Edited, with an Introduction, by WILFRID WARD. 2 vols. Demy 8vo, 21*s.*

WARNER, Francis, M.D.—Lectures on the Anatomy of Movement. Crown 8vo, 4*s.* 6*d.*

WARTER, J. W.—An Old Shropshire Oak. 2 vols. Demy 8vo, 28*s.*

WEDMORE, Frederick.—The Masters of Genre Painting. With Sixteen Illustrations. Post 8vo, 7*s.* 6*d.*

WHITMAN, Sidney.—Conventional Cant: its Results and Remedy. Crown 8vo, 6*s.*

WHITNEY, Prof. William Dwight. — Essentials of English Grammar, for the Use of Schools. Second Edition. Crown 8vo, 3*s.* 6*d.*

WHITWORTH, George Clifford.—An Anglo-Indian Dictionary : a Glossary of Indian Terms used in English, and of such English or other Non-Indian Terms as have obtained special meanings in India. Demy 8vo, cloth, 12*s.*

WILSON, Lieut.-Col. C. T.—The Duke of Berwick, Marshal of France, 1702-1734. Demy 8vo, 15*s.*

WILSON, Mrs. R. F.—The Christian Brothers. Their Origin and Work. With a Sketch of the Life of their Founder, the Ven. JEAN BAPTISTE, de la Salle. Crown 8vo, 6*s.*

WOLTMANN, Dr. Alfred, and WOERMANN, Dr. Karl.—History of Painting. With numerous Illustrations. Medium 8vo. Vol. I. Painting in Antiquity and the Middle Ages. 28*s.* ; bevelled boards, gilt leaves, 30*s.* Vol. II. The Painting of the Renascence. 42*s.* ; bevelled boards, gilt leaves, 45*s.*

YOUMANS, Edward L., M.D.—A Class Book of Chemistry, on the Basis of the New System. With 200 Illustrations. Crown 8vo, 5*s.*

YOUMANS, Eliza A.—First Book of Botany. Designed to Cultivate the Observing Powers of Children. With 300 Engravings. New and Cheaper Edition. Crown 8vo, 2*s.* 6*d.*

YOUNG, Arthur.—Axial Polarity of Man's Word-Embodied Ideas, and its Teaching. Demy 4to, 15*s.*

THE INTERNATIONAL SCIENTIFIC SERIES.

I. **Forms of Water in Clouds and Rivers, Ice and Glaciers.** By J. Tyndall, LL.D., F.R.S. With 25 Illustrations. Ninth Edition. 5*s.*

II. **Physics and Politics**; or, Thoughts on the Application of the Principles of "Natural Selection" and "Inheritance" to Political Society. By Walter Bagehot. Eighth Edition. 4*s.*

III. **Foods.** By Edward Smith, M.D., LL.B., F.R.S. With numerous Illustrations. Ninth Edition. 5*s.*

IV. **Mind and Body**: the Theories and their Relation. By Alexander Bain, LL.D. With Four Illustrations. Eighth Edition. 4*s.*

V. **The Study of Sociology.** By Herbert Spencer. Thirteenth Edition. 5*s.*

VI. **On the Conservation of Energy.** By Balfour Stewart, M.A., LL.D., F.R.S. With 14 Illustrations. Seventh Edition. 5*s.*

VII. **Animal Locomotion**; or Walking, Swimming, and Flying. By J. B. Pettigrew, M.D., F.R.S., etc. With 130 Illustrations. Third Edition. 5*s.*

VIII. **Responsibility in Mental Disease.** By Henry Maudsley, M.D. Fourth Edition. 5*s.*

IX. **The New Chemistry.** By Professor J. P. Cooke. With 31 Illustrations. Ninth Edition. 5*s.*

X. **The Science of Law.** By Professor Sheldon Amos. Sixth Edition. 5*s.*

XI. **Animal Mechanism**: a Treatise on Terrestrial and Aerial Locomotion. By Professor E. J. Marey. With 117 Illustrations. Third Edition. 5*s.*

XII. **The Doctrine of Descent and Darwinism.** By Professor Oscar Schmidt. With 26 Illustrations. Seventh Edition. 5*s.*

XIII. **The History of the Conflict between Religion and Science.** By J. W. Draper, M.D., LL.D. Twentieth Edition. 5*s.*

XIV. **Fungi**: their Nature, Influences, Uses, etc. By M. C. Cooke, M.D., LL.D. Edited by the Rev. M. J. Berkeley, M.A., F.L.S. With numerous Illustrations. Third Edition. 5*s.*

XV. **The Chemical Effects of Light and Photography.** By Dr. Hermann Vogel. With 100 Illustrations. Fourth Edition. 5*s.*

XVI. **The Life and Growth of Language.** By Professor William Dwight Whitney. Fifth Edition. 5*s.*

XVII. **Money and the Mechanism of Exchange.** By W Stanley Jevons, M.A., F.R.S. Eighth Edition. 5*s.*

XVIII. **The Nature of Light.** With a General Account of Physical Optics. By Dr. Eugene Lommel. With 188 Illustrations and a Table of Spectra in Chromo-lithography. Fourth Edition. 5*s.*

XIX. **Animal Parasites and Messmates.** By P. J. Van Beneden. With 83 Illustrations. Third Edition. 5*s.*

XX. **Fermentation.** By Professor Schützenberger. With 28 Illustrations. Fourth Edition. 5*s.*

XXI. **The Five Senses of Man.** By Professor Bernstein. With 91 Illustrations. Fifth Edition. 5*s.*

XXII. **The Theory of Sound in its Relation to Music.** By Professor Pietro Blaserna. With numerous Illustrations. Third Edition. 5*s.*

XXIII. **Studies in Spectrum Analysis.** By J. Norman Lockyer, F.R.S. With six photographic Illustrations of Spectra, and numerous engravings on Wood. Fourth Edition. 6*s.* 6*d.*

XXIV. **A History of the Growth of the Steam Engine.** By Professor R. H. Thurston. With numerous Illustrations. Fourth Edition. 6*s.* 6*d.*

XXV. **Education as a Science.** By Alexander Bain, LL.D. Sixth Edition. 5*s.*

XXVI. **The Human Species.** By Professor A. de Quatrefages. Fourth Edition. 5*s.*

XXVII. **Modern Chromatics.** With Applications to Art and Industry. By Ogden N. Rood. With 130 original Illustrations. Second Edition. 5*s.*

XXVIII. **The Crayfish :** an Introduction to the Study of Zoology. By Professor T. H. Huxley. With 82 Illustrations. Fourth Edition. 5*s.*

XXIX. **The Brain as an Organ of Mind.** By H. Charlton Bastian, M.D. With numerous Illustrations. Third Edition. 5*s.*

XXX. **The Atomic Theory.** By Prof. Wurtz. Translated by G. Cleminshaw, F.C.S. Fourth Edition. 5*s.*

XXXI. **The Natural Conditions of Existence as they affect Animal Life.** By Karl Semper. With 2 Maps and 106 Woodcuts. Third Edition. 5*s.*

XXXII. **General Physiology of Muscles and Nerves.** By Prof. J. Rosenthal. Third Edition. With Illustrations. 5*s.*

XXXIII. **Sight :** an Exposition of the Principles of Monocular and Binocular Vision. By Joseph le Conte, LL.D. Second Edition. With 132 Illustrations. 5*s.*

XXXIV. **Illusions:** a Psychological Study. By James Sully. Third Edition. 5*s.*

XXXV. **Volcanoes: what they are and what they teach.** By Professor J. W. Judd, F.R.S. With 92 Illustrations on Wood. Third Edition. 5*s.*

XXXVI. **Suicide:** an Essay on Comparative Moral Statistics. By Prof. H. Morselli. Second Edition. With Diagrams. 5*s.*

XXXVII. **The Brain and its Functions.** By J. Luys. With Illustrations. Second Edition. 5*s.*

XXXVIII. **Myth and Science:** an Essay. By Tito Vignoli. Third Edition. 5*s.*

XXXIX. **The Sun.** By Professor Young. With Illustrations. Second Edition. 5*s.*

XL. **Ants, Bees, and Wasps:** a Record of Observations on the Habits of the Social Hymenoptera. By Sir John Lubbock, Bart., M.P. With 5 Chromo-lithographic Illustrations. Eighth Edition. 5*s.*

XLI. **Animal Intelligence.** By G. J. Romanes, LL.D., F.R.S. Fourth Edition. 5*s.*

XLII. **The Concepts and Theories of Modern Physics.** By J. B. Stallo. Third Edition. 5*s.*

XLIII. **Diseases of the Memory;** An Essay in the Positive Psychology. By Prof. Th. Ribot. Third Edition. 5*s.*

XLIV. **Man before Metals.** By N. Joly, with 148 Illustrations. Fourth Edition. 5*s.*

XLV. **The Science of Politics.** By Prof. Sheldon Amos. Third Edition. 5*s.*

XLVI. **Elementary Meteorology.** By Robert H. Scott. Fourth Edition. With Numerous Illustrations. 5*s.*

XLVII. **The Organs of Speech and their Application in the Formation of Articulate Sounds.** By Georg Hermann Von Meyer. With 47 Woodcuts. 5*s.*

XLVIII. **Fallacies.** A View of Logic from the Practical Side. By Alfred Sidgwick. Second Edition. 5*s.*

XLIX. **Origin of Cultivated Plants.** By Alphonse de Candolle. 5*s.*

L. **Jelly-Fish, Star-Fish, and Sea-Urchins.** Being a Research on Primitive Nervous Systems. By G. J. Romanes. With Illustrations. 5*s.*

LI. **The Common Sense of the Exact Sciences.** By the late William Kingdon Clifford. Second Edition. With 100 Figures. 5*s.*

LII. **Physical Expression: Its Modes and Principles.** By Francis Warner, M.D., F.R.C.P., Hunterian Professor of Comparative Anatomy and ¡Physiology, R.C.S.E. With 50 Illustrations. 5*s*.

LIII. **Anthropoid Apes.** By Robert Hartmann. With 63 Illustrations. 5*s*.

LIV. **The Mammalia in their Relation to Primeval Times.** By Oscar Schmidt. With 51 Woodcuts. 5*s*.

LV. **Comparative Literature.** By H. Macaulay Posnett, LL.D. 5*s*.

LVI. **Earthquakes and other Earth Movements.** By Prof. John Milne. With 38 Figures. Second Edition. 5*s*.

LVII. **Microbes, Ferments, and Moulds.** By E. L. Trouessart. With 107 Illustrations. 5*s*.

LVIII. **Geographical and Geological Distribution of Animals.** By Professor A. Heilprin. With Frontispiece. 5*s*.

LIX. **Weather.** A Popular Exposition of the Nature of Weather Changes from Day to Day. By the Hon. Ralph Abercromby. With 96 Illustrations. 5*s*.

LX. **Animal Magnetism.** By Alfred Binet and Charles Féré. 5*s*.

LXI. **Manual of British Discomycetes,** with descriptions of all the Species of Fungi hitherto found in Britain included in the Family, and Illustrations of the Genera. By William Phillips, F.L.S. 5*s*.

LXII. **International Law.** With Materials for a Code of International Law. By Professor Leone Levi. 5*s*.

LXIII. **The Origin of Floral Structures through Insect Agency.** By Prof. G. Henslow.

MILITARY WORKS.

BRACKENBURY, Col. C. B., R.A. — Military Handbooks for Regimental Officers.

    I. **Military Sketching and Reconnaissance.** By Col. F. J. Hutchison and Major H. G. MacGregor. Fifth Edition. With 15 Plates. Small crown 8vo, 4*s*.

    II. **The Elements of Modern Tactics Practically applied to English Formations.** By Lieut.-Col. Wilkinson Shaw. Sixth Edition. With 25 Plates and Maps. Small crown 8vo, 9*s*.

    III. **Field Artillery.** Its Equipment, Organization and Tactics. By Major Sisson C. Pratt, R.A. With 12 Plates. Third Edition. Small crown 8vo, 6*s*.

D

BRACKENBURY, Col. C. B., R.A.—continued.

IV. The Elements of Military Administration. First Part: Permanent System of Administration. By Major J. W. Buxton. Small crown 8vo, 7s. 6d.

V. Military Law: Its Procedure and Practice. By Major Sisson C. Pratt, R.A. Third Edition. Small crown 8vo, 4s. 6d.

VI. Cavalry in Modern War. By Col. F. Chenevix Trench. Small crown 8vo, 6s.

VII. Field Works. Their Technical Construction and Tactical Application. By the Editor, Col. C. B. Brackenbury, R.A. Small crown 8vo.

BRENT, Brig.-Gen. J. L.—Mobilizable Fortifications and their Controlling Influence in War. Crown 8vo, 5s.

BROOKE, Major, C. K.—A System of Field Training. Small crown 8vo, cloth limp, 2s.

Campaign of Fredericksburg, November—December, 1862. A Study for Officers of Volunteers. With 5 Maps and Plans. Crown 8vo, 5s.

CLERY, C., Lieut.-Col.—Minor Tactics. With 26 Maps and Plans. Seventh Edition, Revised. Crown 8vo, 9s.

COLVILE, Lieut. Col. C. F.—Military Tribunals. Sewed, 2s. 6d.

CRAUFURD, Capt. H. J.—Suggestions for the Military Training of a Company of Infantry. Crown 8vo, 1s. 6d.

HAMILTON, Capt. Ian, A.D.C.—The Fighting of the Future. 1s.

HARRISON, Col. R.—The Officer's Memorandum Book for Peace and War. Fourth Edition, Revised throughout. Oblong 32mo, red basil, with pencil, 3s. 6d.

Notes on Cavalry Tactics, Organisation, etc. By a Cavalry Officer. With Diagrams. Demy 8vo, 12s.

PARR, Capt. H. Hallam, C.M.G.—The Dress, Horses, and Equipment of Infantry and Staff Officers. Crown 8vo, 1s.

SCHAW, Col. H.—The Defence and Attack of Positions and Localities. Third Edition, Revised and Corrected. Crown 8vo, 3s. 6d.

STONE, Capt. F. Gleadowe, R.A.—Tactical Studies from the Franco–German War of 1870-71. With 22 Lithographic Sketches and Maps. Demy 8vo, 30s.

WILKINSON, H. Spenser, Capt. 20th Lancashire R.V.—Citizen Soldiers. Essays towards the Improvement of the Volunteer Force. Crown 8vo, 2s. 6d.

POETRY.

ABBAY, R.—The Castle of Knaresborough. A Tale in Verse. Crown 8vo, 6s.

ADAM OF ST. VICTOR.—The Liturgical Poetry of Adam of St. Victor. From the text of GAUTIER. With Translations into English in the Original Metres, and Short Explanatory Notes, by DIGBY S. WRANGHAM, M.A. 3 vols. Crown 8vo, printed on hand-made paper, boards, 21s.

AITCHISON, James.—The Chronicle of Mites. A Satire. Small crown 8vo. 5s.

ALEXANDER, William, D.D., Bishop of Derry.—St. Augustine's Holiday, and other Poems. Crown 8vo, 6s.

AUCHMUTY, A. C.—Poems of English Heroism : From Brunan-burh to Lucknow ; from Athelstan to Albert. Small crown 8vo, 1s. 6d.

BARNES, William.—Poems of Rural Life, in the Dorset Dialect. New Edition, complete in one vol. Crown 8vo, 8s. 6d.

BAYNES, Rev. Canon H. R.—Home Songs for Quiet Hours. Fourth and Cheaper Edition. Fcap. 8vo, cloth, 2s. 6d.

BEVINGTON, L. S.—Key Notes. Small crown 8vo, 5s.

BLUNT, Wilfrid Scawen. — The Wind and the Whirlwind. Demy 8vo, 1s. 6d.

The Love Sonnets of Proteus. Fifth Edition, 18mo. Cloth extra, gilt top, 5s.

BOWEN, H. C., M.A.—Simple English Poems. English Literature for Junior Classes. In Four Parts. Parts I., II., and III., 6d. each, and Part IV., 1s. Complete, 3s.

BRYANT, W. C.—Poems. Cheap Edition, with Frontispiece. Small crown 8vo, 3s. 6d.

Calderon's Dramas : the Wonder-Working Magician — Life is a Dream—the Purgatory of St. Patrick. Translated by DENIS FLORENCE MACCARTHY. Post 8vo, 10s.

Camoens' Lusiads. — Portuguese Text, with Translation by J. J. AUBERTIN. Second Edition. 2 vols. Crown 8vo, 12s.

CAMPBELL, Lewis.—Sophocles. The Seven Plays in English Verse. Crown 8vo, 7s. 6d.

CERVANTES.—Journey to Parnassus. Spanish Text, with Translation into English Tercets, Preface, and Illustrative Notes, by JAMES Y. GIBSON. Crown 8vo, 12s.

CERVANTES—continued.

Numantia: a Tragedy. Translated from the Spanish, with Introduction and Notes, by JAMES Y. GIBSON. Crown 8vo, printed on hand-made paper, 5*s.*

Chronicles of Christopher Columbus. A Poem in 12 Cantos. By M. D. C. Crown 8vo, 7*s.* 6*d.*

Cid Ballads, and other Poems.—Translated from Spanish and German by J. Y. GIBSON. 2 vols. Crown 8vo, 12*s.*

COXHEAD, Ethel.—Birds and Babies. With 33 Illustrations. Imp. 16mo, gilt, 2*s.* 6*d.*

Dante's Divina Commedia. Translated in the *Terza Rima* of Original, by F. K. H. HASELFOOT. Demy 8vo, 16*s.*

DE BERANGER.—A Selection from his Songs. In English Verse. By WILLIAM TOYNBEE. Small crown 8vo, 2*s.* 6*d.*

DENNIS, J.—English Sonnets. Collected and Arranged by. Small crown 8vo, 2*s.* 6*d.*

DE VERE, Aubrey.—Poetical Works.

    I. THE SEARCH AFTER PROSERPINE, etc. 6*s.*
    II. THE LEGENDS OF ST. PATRICK, etc. 6*s.*
    III. ALEXANDER THE GREAT, etc. 6*s.*

The Foray of Queen Meave, and other Legends of Ireland's Heroic Age. Small crown 8vo, 5*s.*

Legends of the Saxon Saints. Small crown 8vo, 6*s.*

Legends and Records of the Church and the Empire. Small crown 8vo, 6*s.*

DILLON, Arthur.—Gods and Men. Fcap. 4to, 7*s.* 6*d.*

DOBSON, Austin.—Old World Idylls and other Verses. Seventh Edition. Elzevir 8vo, gilt top, 6*s.*

At the Sign of the Lyre. Fifth Edition. Elzevir 8vo, gilt top, 6*s.*

DOWDEN, Edward, LL.D.—Shakspere's Sonnets. With Introduction and Notes. Large post 8vo, 7*s.* 6*d.*

DUTT, Toru.—A Sheaf Gleaned in French Fields. New Edition. Demy 8vo, 10*s.* 6*d.*

Ancient Ballads and Legends of Hindustan. With an Introductory Memoir by EDMUND GOSSE. Second Edition, 18mo. Cloth extra, gilt top, 5*s.*

EDWARDS, Miss Betham.—Poems. Small crown 8vo, 3*s.* 6*d.*

ELLIOTT, Ebenezer, The Corn Law Rhymer.—Poems. Edited by his son, the Rev. EDWIN ELLIOTT, of St. John's, Antigua. 2 vols. Crown 8vo, 18*s.*

English Verse. Edited by W. J. LINTON and R. H. STODDARD.
5 vols. Crown 8vo, cloth, 5*s.* each.
I. CHAUCER TO BURNS.
II. TRANSLATIONS.
III. LYRICS OF THE NINETEENTH CENTURY.
IV. DRAMATIC SCENES AND CHARACTERS.
V. BALLADS AND ROMANCES.

FOSKETT, Edward.—**Poems.** Crown 8vo, 6*s.*

GOODCHILD, John A.—**Somnia Medici.** Three series. Small
crown 8vo, 5*s.* each.

GOSSE, Edmund.—**New Poems.** Crown 8vo, 7*s. 6d.*

    Firdausi in Exile, and other Poems. Second Edition. Elzevir
8vo, gilt top, 6*s.*

GURNEY, Rev. Alfred.—**The Vision of the Eucharist,** and other
Poems. Crown 8vo, 5*s.*

    A Christmas Faggot. Small crown 8vo, 5*s.*

HARRISON, Clifford.—**In Hours of Leisure.** Crown 8vo, 5*s.*

HEYWOOD, J. C.—**Herodias,** a Dramatic Poem. New Edition,
Revised. Small crown 8vo, 5*s.*

    Antonius. A Dramatic Poem. New Edition, Revised. Small
crown 8vo, 5*s.*

    Salome. A Dramatic Poem. Small crown 8vo, 5*s.*

HICKEY, E. H.—**A Sculptor,** and other Poems. Small crown
8vo, 5*s.*

HOLE, W. G.—**Procris,** and other Poems. Fcap. 8vo, 3*s. 6d.*

KEATS, John.—**Poetical Works.** Edited by W. T. ARNOLD. Large
crown 8vo, choicely printed on hand-made paper, with Portrait
in *eau-forte.* Parchment or cloth, 12*s.* ; vellum, 15*s.*

KING, Edward. **A Venetian Lover.** Small 4to, 6*s.*

KING, Mrs. Hamilton.—**The Disciples.** Ninth Edition, and Notes.
Small crown 8vo, 5*s.*

    A Book of Dreams. Second Edition. Crown 8vo, 3*s. 6d.*

LAFFAN, Mrs. R. S. De Courcy.—**A Song of Jubilee,** and other
Poems. With Frontispiece. Small crown 8vo, 3*s. 6d.*

LANG, A.—**XXXII. Ballades in Blue China.** Elzevir 8vo, 5*s.*

    Rhymes à la Mode. With Frontispiece by E. A. Abbey.
Second Edition. Elzevir 8vo, cloth extra, gilt top, 5*s.*

LANGFORD, J. A., LL.D.—**On Sea and Shore.** Small crown
8vo, 5*s.*

LASCELLES, John.—Golden Fetters, and other Poems. Small crown 8vo, 3s. 6d.

LAWSON, Right Hon. Mr. Justice.—Hymni Usitati Latine Redditi : with other Verses. Small 8vo, parchment, 5s.

Living English Poets MDCCCLXXXII. With Frontispiece by Walter Crane. Second Edition. Large crown 8vo. Printed on hand-made paper. Parchment or cloth, 12s. ; vellum, 15s.

LOCKER, F.—London Lyrics. Tenth Edition. With Portrait, Elzevir 8vo. Cloth extra, gilt top, 5s.

Love in Idleness. A Volume of Poems. With an Etching by W. B. Scott. Small crown 8vo, 5s.

LUMSDEN, Lieut.-Col. H. W.—Beowulf : an Old English Poem. Translated into Modern Rhymes. Second and Revised Edition. Small crown 8vo, 5s.

LYSAGHT, Sidney Royse.—A Modern Ideal. A Dramatic Poem. Small crown 8vo, 5s.

MAGNUSSON, Eirikr, M.A., and PALMER, E. H., M.A.—Johan Ludvig Runeberg's Lyrical Songs, Idylls, and Epigrams. Fcap. 8vo, 5s.

MEREDITH, Owen [The Earl of Lytton].—Lucile. New Edition. With 32 Illustrations. 16mo, 3s. 6d. Cloth extra, gilt edges, 4s. 6d.

MORRIS, Lewis.—Poetical Works of. New and Cheaper Editions, with Portrait. Complete in 3 vols., 5s. each.
Vol. I. contains "Songs of Two Worlds." Twelfth Edition.
Vol. II. contains "The Epic of Hades." Twenty-first Edition.
Vol. III. contains "Gwen" and "The Ode of Life." Seventh Edition.
Vol. IV. contains "Songs Unsung" and "Gycia." Fifth Edition.

Songs of Britain. Third Edition. Fcap. 8vo, 5s.

The Epic of Hades. With 16 Autotype Illustrations, after the Drawings of the late George R. Chapman. 4to, cloth extra, gilt leaves, 21s.

The Epic of Hades. Presentation Edition. 4to, cloth extra, gilt leaves, 10s. 6d.

The Lewis Morris Birthday Book. Edited by S. S. COPEMAN, with Frontispiece after a Design by the late George R. Chapman. 32mo, cloth extra, gilt edges, 2s. ; cloth limp, 1s. 6d.

MORSHEAD, E. D. A.—The House of Atreus. Being the Agamemnon, Libation-Bearers, and Furies of Æschylus. Translated into English Verse. Crown 8vo, 7s.

The Suppliant Maidens of Æschylus. Crown 8vo, 3s. 6d.

MOZLEY, J. Rickards.—The Romance of Dennell. A Poem in Five Cantos. Crown 8vo, 7s. 6d.

MULHOLLAND, Rosa.—Vagrant Verses. Small crown 8vo, 5s.

NADEN, Constance C. W.—A Modern Apostle, and other Poems. Small crown 8vo, 5s.

NOEL, The Hon. Roden.—A Little Child's Monument. Third Edition. Small crown 8vo, 3s. 6d.

The House of Ravensburg. New Edition. Small crown 8vo, 6s.

The Red Flag, and other Poems. New Edition. Small crown 8vo, 6s.

Songs of the Heights and Deeps. Crown 8vo, 6s.

O'BRIEN, Charlotte Grace.—Lyrics. Small crown 8vo, 3s. 6d.

O'HAGAN, John.—The Song of Roland. Translated into English Verse. New and Cheaper Edition. Crown 8vo, 5s.

PFEIFFER, Emily.—The Rhyme of the Lady of the Rock, and How it Grew. Second Edition. Small crown 8vo, 3s. 6d.

Gerard's Monument, and other Poems. Second Edition. Crown 8vo, 6s.

Under the Aspens: Lyrical and Dramatic. With Portrait. Crown 8vo, 6s.

PIATT, J. J.—Idyls and Lyrics of the Ohio Valley. Crown 8vo, 5s.

PREVOST, Francis.—Melilot. 3s. 6d.

Fires of Green Wood. Small crown 8vo, 3s. 6d.

Rare Poems of the 16th and 17th Centuries. Edited by W. J. LINTON. Crown 8vo, 5s.

RHOADES, James.—The Georgics of Virgil. Translated into English Verse. Small crown 8vo, 5s.

Poems. Small crown 8vo, 4s. 6d.

Dux Redux. A Forest Tangle. Small crown 8vo, 3s. 6d.

ROBINSON, A. Mary F.—A Handful of Honeysuckle. Fcap. 8vo, 3s. 6d.

The Crowned Hippolytus. Translated from Euripides. With New Poems. Small crown 8vo, 5s.

SCHILLER, Friedrich.—Wallenstein. A Drama. Done in English Verse, by J. A. W. HUNTER, M.A. Crown 8vo, 7s. 6d.

SCHWARTZ, J. M. W.—Nivalis. A Tragedy in Five Acts. Small crown 8vo, 5s.

SCOTT, E. J. L.—The Eclogues of Virgil.—Translated into English Verse. Small crown 8vo, 3*s.* 6*d.*

SHERBROOKE, Viscount.—Poems of a Life. Second Edition. Small crown 8vo, 2*s.* 6*d.*

SINCLAIR, Julian.—Nakiketas, and other Poems. Small crown 8vo, 2*s.* 6*d.*

SMITH, J. W. Gilbart.—The Loves of Vandyck. A Tale of Genoa. Small crown 8vo, 2*s.* 6*d.*

The Log o' the "Norseman." Small crown 8vo, 5*s.*

Serbelloni. Small crown 8vo, 5*s.*

Sophocles : The Seven Plays in English Verse. Translated by LEWIS CAMPBELL. Crown 8vo, 7*s.* 6*d.*

STEWART, Phillips.—Poems. Small crown 8vo, 2*s.* 6*d.*

SYMONDS, John Addington.—Vagabunduli Libellus. Crown 8vo, 6*s.*

Tasso's Jerusalem Delivered. Translated by Sir JOHN KINGSTON JAMES, Bart. Two Volumes. Printed on hand-made paper, parchment, bevelled boards. Large crown 8vo, 21*s.*

TAYLOR, Sir H.—Works. Complete in Five Volumes. Crown 8vo, 30*s.*

Philip Van Artevelde. Fcap. 8vo, 3*s.* 6*d.*

The Virgin Widow, etc. Fcap. 8vo, 3*s.* 6*d.*

The Statesman. Fcap. 8vo, 3*s.* 6*d.*

TODHUNTER, Dr. J.—Laurella, and other Poems. Crown 8vo, 6*s.* 6*d.*

Forest Songs. Small crown 8vo, 3*s.* 6*d.*

The True Tragedy of Rienzi : a Drama. 3*s.* 6*d.*

Alcestis : a Dramatic Poem. Extra fcap. 8vo, 5*s.*

Helena in Troas. Small crown 8vo, 2*s.* 6*d.*

TOMKINS, Zitella E.—Sister Lucetta, and other Poems. Small crown 8vo, 3*s.* 6*d.*

TYNAN, Katherine.—Louise de la Valliere, and other Poems. Small crown 8vo, 3*s.* 6*d.*

Shamrocks. Small crown 8vo, 5*s.*

Unspoken Thoughts. Small crown 8vo, 3*s.* 6*d.*

Victorian Hymns : English Sacred Songs of Fifty Years. Dedicated to the Queen. Large post 8vo, 10*s.* 6*d.*

WEBSTER, Augusta.—In a Day : a Drama. Small crown 8vo, 2*s.* 6*d.*

Disguises : a Drama. Small crown 8vo, 5*s.*

WILLIAMS, James.—A Lawyer's Leisure. Small crown 8vo, 3*s.* 6*d.*

WOOD, Edmund.—Poems. Small crown 8vo, 3*s*. 6*d*.

Wordsworth Birthday Book, The. Edited by ADELAIDE and VIOLET WORDSWORTH. 32mo, limp cloth, 1*s*. 6*d*. ; cloth extra, 2*s*.

YOUNGS, Ella Sharpe.—Paphus, and other Poems. Small crown 8vo, 3*s*. 6*d*.

    A Heart's Life, Sarpedon, and other Poems. Small crown 8vo, 5*s*. 6*d*.

    The Apotheosis of Antinous, and other Poems. With Portrait. Small crown 8vo, 10*s*. 6*d*.

NOVELS AND TALES.

" All But : " a Chronicle of Laxenford Life. By PEN OLIVER, F.R.C.S. With 20 Illustrations. Second Edition. Crown 8vo, 6*s*.

BANKS, Mrs. G. L.—God's Providence House. New Edition. Crown 8vo, 3*s*. 6*d*.

CHICHELE, Mary.—Doing and Undoing. A Story. Crown 8vo, 4*s*. 6*d*.

Danish Parsonage. By an Angler. Crown 8vo, 6*s*.

GRAY, Maxwell. — The Silence of Dean Maitland. Fifth Edition. With Frontispiece. Crown 8vo, 6*s*.

HUNTER, Hay.—The Crime of Christmas Day. A Tale of the Latin Quarter. By the Author of "My Ducats and my Daughter." 1*s*.

HUNTER, Hay, and WHYTE, Walter.—My Ducats and My Daughter. New and Cheaper Edition. With Frontispiece. Crown 8vo, 6*s*.

INGELOW, Jean.—Off the Skelligs : a Novel. With Frontispiece. Second Edition. Crown 8vo, 6*s*.

JENKINS, Edward.—A Secret of Two Lives. Crown 8vo, 2*s*. 6*d*.

KIELLAND, Alexander L.—Garman and Worse. A Norwegian Novel. Authorized Translation, by W. W. Kettlewell. Crown 8vo, 6*s*.

LANG, Andrew.—In the Wrong Paradise, and other Stories. Second Edition. Crown 8vo, 6*s*.

MACDONALD, G.—Donal Grant. A Novel. Second Edition. With Frontispiece. Crown 8vo, 6*s*.

    Home Again. With Frontispiece. Crown 8vo, 6*s*.

    Castle Warlock. A Novel. Second Edition. With Frontispiece. Crown 8vo, 6*s*.

MACDONALD, G.—continued.

Malcolm. With Portrait of the Author engraved on Steel. Eighth Edition. Crown 8vo, 6s.

The Marquis of Lossie. Seventh Edition. With Frontispiece. Crown 8vo, 6s.

St. George and St. Michael. Fifth Edition. With Frontispiece. Crown 8vo, 6s.

What's Mine's Mine. Second Edition. With Frontispiece. Crown 8vo, 6s.

Annals of a Quiet Neighbourhood. Sixth Edition. With Frontispiece. Crown 8vo, 6s.

The Seaboard Parish : a Sequel to "Annals of a Quiet Neighbourhood." Fourth Edition. With Frontispiece. Crown 8vo, 6s.

Wilfred Cumbermede. An Autobiographical Story. Fourth Edition. With Frontispiece. Crown 8vo, 6s.

Thomas Wingfold, Curate. Fourth Edition. With Frontispiece. Crown 8vo, 6s.

Paul Faber, Surgeon. Fourth Edition. With Frontispiece. Crown 8vo, 6s.

MALET, Lucas.—**Colonel Enderby's Wife.** A Novel. New and Cheaper Edition. With Frontispiece. Crown 8vo, 6s.

MULHOLLAND, Rosa.—**Marcella Grace:** An Irish Novel. Crown 8vo, 6s.

PALGRAVE, W. Gifford.—**Hermann Agha:** an Eastern Narrative. Third Edition. Crown 8vo, 6s.

SHAW, Flora L.—**Castle Blair :** a Story of Youthful Days. New and Cheaper Edition. Crown 8vo, 3s. 6d.

STRETTON, Hesba.—**Through a Needle's Eye :** a Story. New and Cheaper Edition, with Frontispiece. Crown 8vo, 6s.

TAYLOR, Col. Meadows, C.S.I., M.R.I.A.—**Seeta :** a Novel. With Frontispiece. Crown 8vo, 6s.

Tippoo Sultaun : a Tale of the Mysore War. With Frontispiece. Crown 8vo, 6s.

Ralph Darnell. With Frontispiece. Crown 8vo, 6s.

A Noble Queen. With Frontispiece. Crown 8vo, 6s.

The Confessions of a Thug. With Frontispiece. Crown 8vo, 6s.

Tara : a Mahratta Tale. With Frontispiece. Crown 8vo, 6s.

Within Sound of the Sea. With Frontispiece. Crown 8vo, 6s.

BOOKS FOR THE YOUNG.

Brave Men's Footsteps. A Book of Example and Anecdote for Young People. By the Editor of "Men who have Risen." With 4 Illustrations by C. Doyle. Ninth Edition. Crown 8vo, 3*s*. 6*d*.

COXHEAD, Ethel.—**Birds and Babies.** With 33 Illustrations. Second Edition. Imp. 16mo, cloth gilt, 2*s*. 6*d*.

DAVIES, G. Christopher.—**Rambles and Adventures of our School Field Club.** With 4 Illustrations. New and Cheaper Edition. Crown 8vo, 3*s*. 6*d*.

EDMONDS, Herbert.—**Well Spent Lives :** a Series of Modern Biographies. New and Cheaper Edition. Crown 8vo, 3*s*. 6*d*.

EVANS, Mark.—**The Story of our Father's Love,** told to Children. Sixth and Cheaper Edition of Theology for Children. With 4 Illustrations. Fcap. 8vo, 1*s*. 6*d*.

MAC KENNA, S. J.—**Plucky Fellows.** A Book for Boys. With 6 Illustrations. Fifth Edition. Crown 8vo, 3*s*. 6*d*.

MALET, Lucas.—**Little Peter.** A Christmas Morality for Children of any Age. With numerous Illustrations. 5*s*.

REANEY, Mrs. G. S.—**Waking and Working ;** or, From Girlhood to Womanhood. New and Cheaper Edition. With a Frontispiece. Crown 8vo, 3*s*. 6*d*.

Blessing and Blessed : a Sketch of Girl Life. New and Cheaper Edition. Crown 8vo, 3*s*. 6*d*.

Rose Gurney's Discovery. A Story for Girls. Dedicated to their Mothers. Crown 8vo, 3*s*. 6*d*.

English Girls : Their Place and Power. With Preface by the Rev. R. W. Dale. Fifth Edition. Fcap. 8vo, 2*s*. 6*d*.

Just Anyone, and other Stories. Three Illustrations. Royal 16mo, 1*s*. 6*d*.

Sunbeam Willie, and other Stories. Three Illustrations. Royal 16mo, 1*s*. 6*d*.

Sunshine Jenny, and other Stories. Three Illustrations. Royal 16mo, 1*s*. 6*d*.

STORR, Francis, and TURNER, Hawes.—**Canterbury Chimes ;** or, Chaucer Tales re-told to Children. With 6 Illustrations from the Ellesmere Manuscript. Third Edition. Fcap. 8vo, 3*s*. 6*d*.

STRETTON, Hesba.—**David Lloyd's Last Will.** With 4 Illustrations. New Edition. Royal 16mo, 2*s*. 6*d*.

WHITAKER, Florence.—**Christy's Inheritance.** A London Story. Illustrated. Royal 16mo, 1*s*. 6*d*.

PRINTED BY WILLIAM CLOWES AND SONS, LIMITED,
LONDON AND BECCLES.

· MESSRS.

KEGAN PAUL, TRENCH & CO.'S

EDITIONS OF

SHAKSPERE'S WORKS.

THE PARCHMENT LIBRARY EDITION.

THE AVON EDITION.

The Text of these Editions is mainly that of Delius. Wherever a variant reading is adopted, some good and recognized Shaksperian Critic has been followed. In no case is a new rendering of the text proposed; nor has it been thought necessary to distract the reader's attention by notes or comments.

ARBOR SCIENTIÆ
ARBOR VITÆ

1, PATERNOSTER SQUARE.

[P. T. O.

SHAKSPERE'S WORKS.

THE AVON EDITION.

Printed on thin opaque paper, and forming 12 handy volumes, cloth, 18s., or bound in 6 volumes, 15s.

The set of 12 volumes may also be had in a cloth box, price 21s., or bound in Roan, Persian, Crushed Persian Levant, Calf, or Morocco, and enclosed in an attractive leather box at prices from 31s. 6d. upwards.

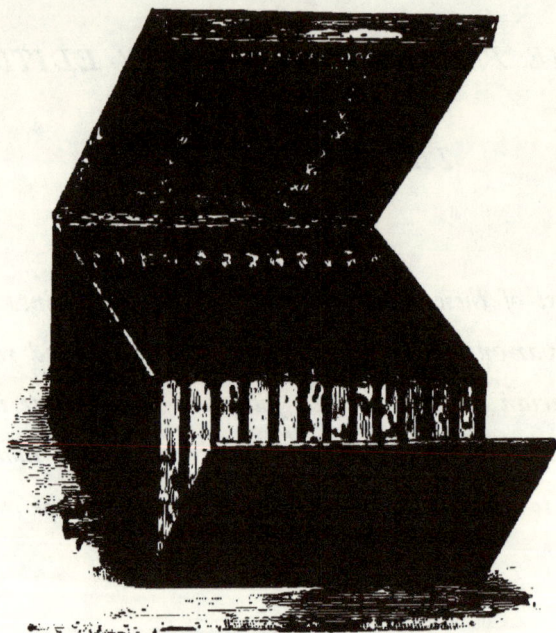

SOME PRESS NOTICES.

"This edition will be useful to those who want a good text, well and clearly printed, in convenient little volumes that will slip easily into an overcoat pocket or a travelling-bag."—*St. James's Gazette.*

"We know no prettier edition of Shakspere for the price."—*Academy.*

"It is refreshing to meet with an edition of Shakspere of convenient size and low price, without either notes or introductions of any sort to distract the attention of the reader."—*Saturday Review.*

"It is exquisite. Each volume is handy, is beautifully printed, and in every way lends itself to the taste of the cultivated student of Shakspere."—*Scotsman.*

LONDON: KEGAN PAUL, TRENCH & CO., 1, PATERNOSTER SQUARE.

SHAKSPERE'S WORKS.

THE PARCHMENT LIBRARY EDITION.

In 12 volumes Elzevir 8vo., choicely printed on hand-made paper, and bound in parchment or cloth, price £3 12s., or in vellum, price £4 10s.

The set of 12 volumes may also be had in a strong cloth box, price £3 17s., or with an oak hanging shelf, £3 18s.

SOME PRESS NOTICES.

Just published.   Price 5s.

AN INDEX TO THE WORKS OF SHAKSPERE.

Applicable to all editions of Shakspere, and giving reference, by topics, to notable passages and significant expressions; brief histories of the plays; geographical names and historic incidents; mention of all characters and sketches of important ones; together with explanations of allusions and obscure and obsolete words and phrases.

By EVANGELINE M. O'CONNOR.

LONDON : KEGAN PAUL, TRENCH & CO., 1, PATERNOSTER SQUARE.

SHAKSPERE'S WORKS.

SPECIMEN OF TYPE.

Salar. My wind, cooling my broth,
Would blow me to an ague, when I thought
What harm a wind too great might do at sea.
I should not see the sandy hour-glass run
But I should think of shallows and of flats,
And see my wealthy Andrew, dock'd in sand,
Vailing her high-top lower than her ribs
To kiss her burial. Should I go to church
And see the holy edifice of stone,
And not bethink me straight of dangerous rocks,
Which touching but my gentle vessel's side,
Would scatter all her spices on the stream,
Enrobe the roaring waters with my silks,
And, in a word, but even now worth this,
And now worth nothing? Shall I have the thought
To think on this, and shall I lack the thought
That such a thing bechanc'd would make me sad?
But tell not me : I know Antonio
Is sad to think upon his merchandise.

Ant. Believe me, no : I thank my fortune for it,
My ventures are not in one bottom trusted,
Nor to one place ; nor is my whole estate
Upon the fortune of this present year :
Therefore my merchandise makes me not sad.

Salar. Why, then you are in love.

Ant. Fie, fie !

Salar. Not in love neither ? Then let us say you
are sad,
Because you are not merry ; and 'twere as easy
For you to laugh, and leap, and say you are merry,
Because you are not sad. Now, by two-headed
Janus,
Nature hath fram'd strange fellows in her time :
Some that will evermore peep through their eyes
And laugh like parrots at a bag-piper ;
And other of such vinegar aspect

LONDON : KEGAN PAUL, TRENCH & CO., 1, PATERNOSTER SQUARE.